A GUIDE
TO BAMBOOS GROWN IN BRITAIN

C.S. Chao

© Royal Botanic Gardens, Kew 1989

First published 1989

General editor of the series M.J.E. Coode.
Special editor of this number T.A. Cope.
English text revised by W.D. Clayton with
the assistance of S.A. Renvoize. Cover
design by Jackie Panter. Set by Pam Arnold,
Christine Beard, Brenda Carey, Margaret
Newman, Helen O'Brien and Pam Rosen.
Illustrations prepared by
Soejatmi Dransfield.

ISBN 0 947643 17 6

PREFACE

Prof. Chao of Nanjing Forestry University, China, spent a year working on the collections at Kew. Part of his brief was to compare Chinese bamboos with horticultural specimens brought to Europe by 19th century plant explorers. These introductions were often new to science at the time, and were named from garden clones without reference to the wild populations. When added to uncertainties over generic delimitation, the result has been a confusion of nomenclature which has persisted to the present day.

Prof. Chao has performed a valuable service by applying his knowledge of native Chinese bamboos to the problem, and presenting his conclusions in this handbook.

Professor G.T. Prance
Director
Royal Botanic Gardens, Kew

INTRODUCTION

Bamboos are a natural group of the grass family, comprising about 50 genera and over 800 species. They are native to Asia (26 genera, more than 500 species), America (19 genera, 260 species) and Africa including Madagascar (13 genera, 30 species). None is native to Europe.

In Asia, bamboos constitute an important forest resource and are commonly used as a substitute for wood in all kinds of artefacts. They are also cultivated as ornamental plants in gardens, and have attracted the attention of European plant collectors who introduced them to European horticulture. It is believed that the first bamboo grown in Britain was *Phyllostachys nigra* in or before 1827. Since then many hardy bamboos have been introduced to the country, and a few have escaped from cultivation and become naturalized.

The classification of bamboos is particularly difficult. This is because the long interval between flowering periods means that floral characters are of no practical value for the identification of garden plants. *Sinarundinaria nitida*, for example, has never been seen in flower since its introduction to Kew in the last century. Distinguishing features must therefore be sought among the vegetative parts, but these often vary considerably between culm and crown branches, or between mature and immature shoots. The keys and descriptions given here are taken from the mature stage.

In all, 14 genera and 70 species, varieties, cultivars and forms are recognized here as generally cultivated in Britain. However, the private collection of David Crampton, designated the 'National' collection in 1987 by The National Council for the Conservation of Plants and Gardens, comprises more than 150 genera and 200 species.

THE BAMBOO PLANT

The rhizome is of great importance for identification and two kinds can be recognized. A monopodial rhizome grows horizontally, producing vertical culms from its lateral buds. It is long, slender and often aggressively running, so giving rise to openly spaced culms, though these may subsequently form clumps by branching at the basal nodes. A sympodial rhizome starts horizontally but soon turns upwards to become a culm, horizontal growth being continued by a lateral bud. It is short, club-shaped and closely spaced, so forming dense clumps; however, in a few species the neck of the club is unusually long and could be mistaken for a monopodial runner (Fig. 1).

The young culms are clad in culm-sheaths. These are modified leaves with an expanded sheath, which wraps around the culm, and a much diminished blade. The junction between sheath and blade bears a membranous fringe (ligule) and is sometimes flanked by little ear-like projections (auricles) or marked by conspicuous bristles (oral setae). The culm-sheaths usually fall away as the culm matures (Fig. 2).

Branches arise in the axils of culm-sheaths. In some genera they have the peculiar property of precociously rebranching from compacted basal nodes, so that the culm apparently produces several branches at a node. The number is of diagnostic value and is termed the branch-complement (Fig. 3).

The foliage leaves consist of a narrow leaf-sheath and a broad leaf-blade, with a ligule at their junction. They may also bear auricles and oral setae.

The classification of bamboos into genera draws upon inflorescence characters which are merely mentioned here. There are three principal kinds of inflorescence: simple, a raceme or panicle; spatheate, a partially condensed type with spikelets subtended by leafy spathes; and condensed, with the branch system greatly contracted and small glume-like bracts subtending the spikelets (Fig. 4).

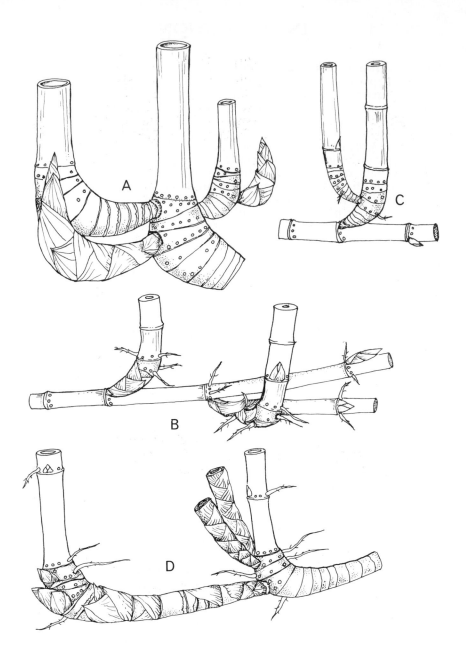

Fig. 1 Rhizome systems. **A** *Bambusa glauscens*. Rhizome short and thicker than the culm it produces; the internodes are very short and condensed. Growth is *sympodial*: new rhizomes grow from lateral buds and turn up at the tip to give rise to new culms (× 0.7). **B** *Pseudosasa japonica*. Rhizome more slender than the culms it produces; internodes very long. Growth is *monopodial*: the rhizome remains below ground and does not turn up at the tip, new culms growing up from lateral buds (× 0.7). **C** *Arundinaria graminea*. Growth is *amphipodial*: a combination of sympodial branches growing from a monopodial rhizome (× 0.7). **D** *Sinarundinaria nitida*. Rhizome thicker than the culm it produces. Growth is *sympodial*: new rhizomes grow from lateral buds; however, the neck of the rhizome is elongated, resulting in widely spaced culms (× 0.5).

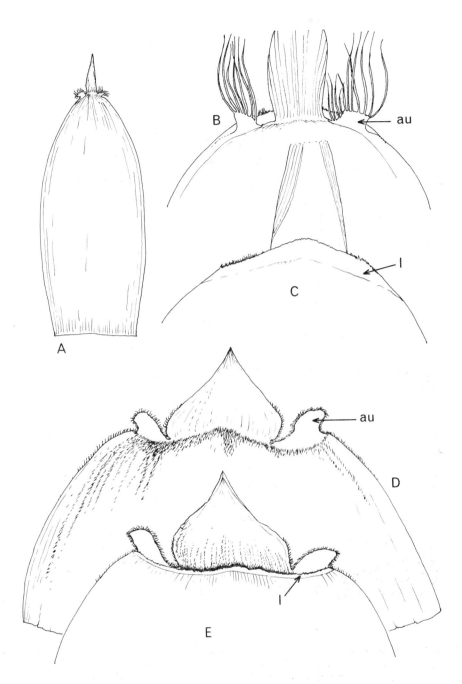

Fig. 2 Culm sheaths: au, auricles; l, ligule. **A–C** *Phyllostachys nigra*. **A**, complete sheath with small strap-shaped blade (× 0.5). **B**, dorsal (outer) view showing detail of the ciliate auricles (× 9). **C**, ventral (inner) view showing detail of the short membranous ligule (× 9). **D & E** *Bambusa vulgaris*. **D**, dorsal (outer) view showing the broad triangular blade and ciliolate auricles (× 0.5). **E**, ventral (inner) view showing detail of the short membranous ligule (× 0.5).

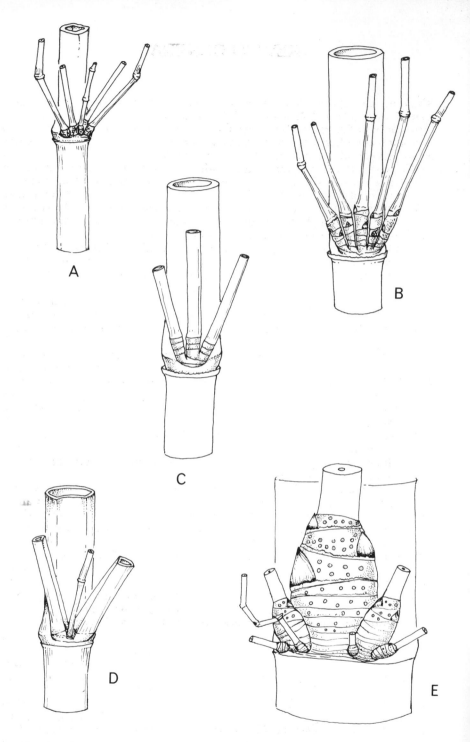

Fig. 3 Branch-complements (all sheaths removed). **A** *Chimonobambusa quad-rangularis*. Three dominant branches and several subsequent minor branches arise at different levels from a single bud; note the square culm (×0.7). **B** *Sinarundinaria nitida*. Several branches of similar size arise at the same level from a single bud (× 3). **C** *Arundinaria graminea*. Three branches of equal size arise at the same level from a single bud (× 3). **D** *Phyllostachys bambusoides*. Branches are paired; if a third branch is present then it is much reduced; note that the culm and branches are flattened on one side (× 0.7). **E** *Bambusa vulgaris*. One branch in the cluster is dominant and much longer than the others; note the swollen base of the branches and the condensed internodes (× 0.7).

KEY TO GENERA

1 Rhizome monopodial with buds and roots from its nodes; culms widely
  spaced, arising from lateral buds of the rhizome: 2
2 Internodes of culm and branches flattened or grooved on one side: 3
  3 Branch-complement typically 2 at each node   **1. Phyllostachys**
  3 Branch-complement usually 3–5 at each node: 4
    4 Culm-sheaths thinly papery, the blades very small and inconspicuous: 5
      5 Branches very short, comprising 1–2 internodes bearing 1 or 2 leaves;
        new shoots emerging in spring   **2. Shibataea**
      5 Branches comprising several internodes bearing 3 or more leaves; new
        shoots emerging in late summer or autumn   **3. Chimonobambusa**
    4 Culm-sheaths coriaceous, the blades conspicuous: 6
      6 Culm-sheaths abscissing completely, their auricles and oral setae strongly
        developed   **4. Sinobambusa**
      6 Culm-sheaths remaining lightly attached at the middle of the base for a
        short time after they dry, their auricles and oral setae lacking or small
               5. Semiarundinaria
2 Internodes of culm and branches cylindrical, at most slightly grooved just
  above the branch-complement; culm-sheaths usually persistent or late-
  falling: 7
  7 Branch-complement 2–3 or more on mid-culm nodes, much more slender
    than culm   **6. Arundinaria**
  7 Branch-complement typically single on mid-culm nodes, equal to the culm
    in diameter: 8
    8 Culms 3–5 m high   **7. Pseudosasa**
    8 Culms less than 3 m high: 9
      9 Culm-sheaths shorter than corresponding internodes; leaves 4–10 on
        each branch   **8. Sasa**
      9 Culm-sheaths equal to or longer than corresponding internodes; leaves
        1–3 on each branch: 10
      10 Leaf-blades less than 4 cm wide, 6-nerved   **8. Sasa**
      10 Leaf-blades very large, 4–8 cm wide, multinerved   **9. Indocalamus**
1 Rhizome sympodial with short neck, or this longer but then without buds or
  roots from its nodes; culms usually caespitose, arising from terminal bud of
  rhizome: 11
11 Culms solid (but if not exceeding 2 m in height see 14. *Bambusa*)
               10. Chusquea
11 Culms hollow: 12
  12 Rhizomes with short or long neck; new shoots emerging in spring; leaf-
    blades with visible tessellation (but a few exceptions); stamens 3:   13
    13 Inflorescence exserted, subtended by small narrow sheaths
               11. Sinarundinaria
    13 Inflorescence enclosed by large spathulate sheaths
               12. Thamnocalamus
  12 Rhizomes with very short neck; new shoots emerging in summer or
    autumn; leaf-blades without visible tessellation; stamens 6:   14
    14 Leaf-blades large, 20–30 cm long, 3–8 cm wide; culms up to 15–20 m
      high   **13. Dendrocalamus**
    14 Leaf-blades smaller, 11–20 cm long, 1–4 cm wide; culms less than 15 m
      high   **14. Bambusa**

Fig. 4 The three principal inflorescence-types. **A** *Pseudosasa japonica*. Simple; a terminal inflorescence often with well developed branches bearing spikelets that are all the same age (whole inflorescence, × 0.7). **B** *Phyllostachys aurea*. Spatheate; the branch-system is partially condensed with the inflorescences axillary and the branches considerably shortened; the spikelets develop at more or less the same time and are subtended by a spathe (a leaf reduced to its sheathing base and occasionally bearing a minute blade) (part of inflorescence, × 0.7). **C** *Bambusa vulgaris*. Condensed; the branch-system is greatly contracted; only small bracts subtending the spikelets remain; growth over a period of time produces dense clusters of spikelets of different ages (part of inflorescence, × 0.7).

DESCRIPTIONS OF GENERA AND SPECIES

1. Phyllostachys Sieb. & Zucc.

Sinoarundinaria Ohwi

Trees or shrubs; rhizomes monopodial with spaced culms. Culm-internodes flattened on one side; branch-complement usually of 2 unequal branches; culm-sheaths deciduous, usually with dark blotches. Inflorescence comprising 1-spikelet racemes gathered into spathulate or capitate fascicles. Spikelets 2–6-flowered without pedicels; glumes 0–3; stamens 3; stigmas 3 on a very long style.

About 55 species. All native to China; a couple of species extending to India, Viet-Nam and Burma; some widely cultivated in Japan. They have been introduced to many countries in the world including America, Europe and Africa. About 20 species in Britain. The private collection of Peter Addington in W. Sussex was designated the 'National' collection in 1987 by the National Council for the Conservation of Plants and Gardens.

Key to species of *Phyllostachys*

1 Culm-sheaths more or less pubescent, usually with auricles and oral setae: 2
  2 Young culms and sheath-scars densely pubescent: 3
    3 Sheath-scars with only one prominent ridge on the lower culm-nodes, two on the upper; culm-sheaths densely black-spotted   **1. P. edulis**
    3 Sheath-scars all with two prominent ridges; culm-sheaths not spotted: 4
      4 Culms wholly or partly black-purple at maturity: 5
        5 Second year culms wholly black-purple   **2a. P. nigra** var. **nigra**
        5 Second year culms partly green: 6
          6 Culms irregularly mottled with black-purple   **2b. P. nigra** f. **punctata**
          6 Culms spotted with black-purple   **2c. P. nigra** f. **boryana**
      4 Culms usually green or grey-green   **2d. P. nigra** var. **henonis**
  2 Young culms and sheath-scars glabrous: 7
    7 Culm-sheaths light yellow, sparsely spotted, their blades strongly wrinkled; leaf-blades densely pubescent beneath   **3. P. dulcis**
    7 Culm-sheaths green-brown to purple-brown, densely spotted; leaf-blades glabrous beneath except for the base: 8
      8 Young culms glaucous, with purple nodes; culm-sheaths with purple nerves; sheath-blades more or less wrinkled   **4. P. viridi-glaucescens**
      8 Young culms not or slightly glaucous, nodes not purple; culm-sheaths without purple nerves; sheath-blades flat: 9
        9 Culms green: 10
          10 Leaf-blades uniformly green   **5a. P. bambusoides** var. **bambusoides**
          10 Leaf-blades green with light green stripes   **5b. P. bambusoides** f. **subvariegata**
        9 Culms yellow with green grooves   **5c. P. bambusoides** var. **castillonii**
1 Culm-sheaths glabrous except for the base: 11
  11 Culm-sheaths with auricles or at least with oral setae: 12
    12 Culm-sheaths green or purplish without spots; leaf-sheaths with erect oral setae: 13
      13 Sheath-blades lanceolate, diverging from the culm and exposing the ligule; leaf-blades densely pubescent beneath   **6. P. bissetii**
      13 Sheath-blades triangular or triangular-lanceolate, closely appressed to the culm and covering the ligule; leaf-blades glabrous
                      7. P. heteroclada
    12 Culm-sheaths not green, more or less spotted; leaf-sheaths with radiate oral setae or the setae deciduous: 14

14 Culm-sheaths densely spotted; young culms glabrous; oral setae of leaf-sheaths conspicuously radiate: 15

15 Culms green: 16

  16 Leaf-blades uniformly green   **5a. P. bambusoides** var. **bambusoides**

  16 Leaf-blades green with light green stripes

                  5b. P. bambusoides f. **subvariegata**

15 Culms yellow with green grooves   **5c. P. bambusoides** var. **castillonii**

14 Culm-sheaths sparsely spotted; young culms pubescent and scabrous; oral setae of leaf-sheaths often deciduous: 17

  17 Culms usually with yellow grooves, the nodes very prominent; ligules of culm-sheaths shortly ciliate at apex   **8. P. aureosulcata**

  17 Culms entirely green, the nodes not prominent; ligules of culm-sheaths bristly   **9. P. mannii**

11 Culm-sheaths without auricles or oral setae: 18

 18 Young culm minutely pubescent along the base line of the culm-sheaths and on the sheath-scars: 19

  19 Basal culm-internodes of normal length   **10. P. meyeri**

  19 Basal culm-internodes usually shortened and swollen: 20

   20 Culms wholly green or yellow-green   **11. P. aurea**

   20 Culms green with yellow grooves **11a. P. aurea** cv. Flavescens-inversa

 18 Young culm completely glabrous: 21

  21 Sheath-scars of lower culm-nodes bearing only one prominent ridge, the upper with two: 22

   22 Culms golden yellow, sometimes with a few green stripes

              12a. P. sulphurea var. **sulphurea**

   22 Culms green or yellowish green   **12b. P. sulphurea** var. **viridis**

  21 Sheath-scars of all culm-nodes with two prominent ridges: 23

   23 Culm-nodes very prominent; culm-wall thick; culm-sheaths scabrous

              13. P. nuda

   23 Culm-nodes scarcely prominent; culm-wall thin; culm-sheaths smooth: 24

   24 Culm-sheaths densely spotted: 25

    25 Ligule of culm-sheath purple-red; sheath-blades flattened

              14. P. makinoi

    25 Ligule of culm-sheath green-brown; sheath-blades strongly wrinkled   **15. P. vivax**

   24 Culm-sheaths sparsely spotted or without spots: 26

    26 Culm-sheaths light green or light yellow, their ligules bristly at the apex: 27

     27 Culm-sheaths light yellow with sparse spots; ligular bristles white

              16. P. angusta

     27 Culm-sheaths light green with red margin, no spots; ligular bristles purple   **17. P. rubromarginata**

    26 Culm-sheaths dark in colour, their ligules ciliate or subglabrous at the apex: 28

     28 Culm-sheaths mealy, their ligules obtuse and pale brown

              18. P. propinqua

     28 Culm-sheaths not mealy, their ligules truncate: 29

      29 Culm-sheaths with short purplish red ligules; leaf-ligules purplish red; internodes not zigzag   **19. P. glauca**

      29 Culm-sheaths with long brown or greenish brown ligules; leaf-ligules not purplish red; lower internodes of some culms zigzag   **20. P. flexuosa**

1. Phyllostachys edulis (*Carr.*) *H. de Lehaie*

Bambusa edulis Carr.
Phyllostachys pubescens H. de Lehaie
P. heterocycla (Carr.) Mitf. var. *pubescens* (H. de Lehaie) Ohwi
P. heterocycla f. *pubescens* (H. de Lehaie) Muroi

English: Moso bamboo, Giant hairy-sheath edible bamboo
Chinese: Mao zhu (Hairy bamboo)
Japanese: Maso-chiku

Culms up to 20–25 m high, 7–30 cm in diameter, green, grey or grey-yellow in colour, marked by one prominent ridge (sheath scar) on the lower culm-nodes, densely pubescent and mealy during the early stage of growth. Culm-sheaths thick, dark, densely dark brown-mottled and pubescent; blades green and wrinkled; auricles small, oral setae conspicuous. Leaf-blades smaller than in any other species of the genus, 4–10 cm long, 4–10 mm wide.

P. edulis is the biggest bamboo of the genus and is native to China, being the principal component in 50% of Chinese bamboo forests. It prefers warmer regions and is widely cultivated in southern China northwards to the Yangtze River, being hardy down to about −6°C to −8°C. It was introduced from China into Japan in 1746, and is believed to have been introduced to France in 1866, but no conclusive evidence has been found.

It is one of the most valuable bamboos of China and Japan, whose culms are used for numerous purposes, such as furniture, implements, paper-making and woven bamboo articles. Commercial production of young edible shoots is also of great economic importance. The spring shoots, though fragrant while cooking, have a somewhat bitter taste, but the winter shoots are excellent and highly esteemed as a delicacy.

This bamboo is rarely grown in Britain, and is not hardy in most gardens.

The species is often erroneously named *P. mitis* in horticultural literature. Another name sometimes found in the literature is *P. edulis* var. *heterocycla* (Carr.) H. de Lehaie (*P. heterocycla, P. pubescens* var. *heterocycla*). This refers to a teratological form whose shortened basal internodes, separated by oblique nodes, bulge grotesquely on alternate sides.

2. Phyllostachys nigra (*Lodd.*) *Munro*

Bambusa nigra Lodd.
Phyllostachys puberula (Miq.) Munro var. *nigra* H. de Lehaie
P. filifera McClure
P. nigripes Hayata

English: Black bamboo
Chinese: Zi zhu (Purple bamboo), Hei zhu (Black bamboo)
Japanese: Gomadake, Kurodake, Shiro-chiku

2a. var. nigra

Culms up to 3–6 m (rarely to 10 m) high, 2–4 cm in diameter, densely covered in pubescence and white powder when young, light green in colour during the first season, gradually developing purplish or purple-black spots and becoming entirely purple-black or black during the second season. Culm-sheaths shorter than internodes, pale pinkish brown, densely brownish-pubescent, without spots; blades light green, broadly triangular in shape and wrinkled; auricles very conspicuous, oblong, purple-black, with long curved oral setae. Leaf-blades small, thin and quite smooth on both surfaces.

P. nigra is native to China. It is widely cultivated from the Yellow River (Beijing) southwards to South China and is a favourite garden ornamental in China and Japan, but it is rare in the wild. The peculiar coloration provided by the older culms, and its hardy habit, brought it to the attention of European horticulturists. It was introduced into Britain in, or possibly before, 1827, and is believed to be the first *Phyllostachys* introduced to Europe.

It, and all its forms, flowered in Europe for the first time in 1900–1905, and again in 1930–1935.

It is an excellent ornamental species, and its culms are of high quality for handicrafts such as furniture, flutes and walking sticks.

2b. f. **punctata** (*Bean*) *Nakai*

Phyllostachys nigra var. *punctata* Bean
P. puberula (Miq.) Munro var. *nigro-punctata* H. de Lehaie
P. nigra cv. Punctata

Like f. *nigra*, but the black coloration broken into irregular blotches.

2c. f. **boryana** (*Mitf.*) *Makino*

Phyllostachys boryana Mitf.
P. puberula (Miq.) Munro var. *boryana* (Mitf.) H. de Lehaie
P. nigra cv. Bory

Like var. *henonis*, but with some brown or dark purple blotches on the otherwise green culms.

2d. var. **henonis** (*Mitf.*) *Rendle*

Phyllostachys henonis Mitf.
P. fauriei Hack.
P. henryi Rendle
P. montana Rendle
P. nevinii Hance
P. nevinii var. *hupehensis* Rendle
P. puberula (Miq.) Munro
P. stauntonii Munro
P. veitchiana Rendle
P. nigra cv. Henon

English: Henon bamboo
Chinese: Mao-jin zhu (Muscle bamboo)
Japanese: Ha-chiku

This bamboo represents the original wild form of *P. nigra*. It is similar to var. *nigra* in most respects but differs in the lack of any colour markings on the green or grey-green culms and in its larger stature, up to 15–20 m high and 8–10 cm in diameter. It is a little misleading that under the rules of nomenclature the wild plant has to be treated as a variety of the earlier described ornamental.

Henon bamboo is a common species in China and always found in the mountains. It was introduced into Europe in 1890 and flowered at the same time as *P. nigra* var. *nigra*. It is very hardy and widely acclaimed as one of loveliest of *Phyllostachys*.

3. Phyllostachys dulcis *McClure*

English: Sweet-shoot bamboo
Chinese: Bai-ke-bu-ji zhu, Bai-bu-ji zhu (White sheath bamboo)

Culms up to 7–8 m high, 4–5 cm in diameter; internodes fairly short (15–24 cm) and glabrous. Culm-sheaths pale yellow with small brownish spots when fresh, sometimes brownish at the margin, sparsely white-hirsute; blades strongly wrinkled; auricles and oral setae very conspicuous. Leaf-blades densely pubescent beneath.

P. dulcis is native to Zhejiang Province of China. It is one of the most important edible species, lacking any bitter flavour, and is widely cultivated in eastern China. The culms are of little industrial value.

It was introduced to the United States in 1908, named by McClure in 1945, and brought to Britain a couple of years ago, growing in a private garden in W. Sussex. Its distribution in Britain is extremely limited.

4. Phyllostachys viridi-glaucescens *(Carr.) A. & C. Riv.*

Bambusa viridi-glaucescens Carr.
Phyllostachys elegans McClure

Chinese: Tian-sun zhu (Sweet-shoot bamboo)

Culms up to 7–8 m high, 4–5 cm in diameter; internodes green and mealy with dark purple nodes when young. Culm-sheaths brownish yellow with numerous purple nerves, usually spotted and hirsute, or the smaller ones subglabrous; blades narrow, more or less wrinkled; auricles and oral setae conspicuous.

This species much resembles *P. bambusoides*, but it has mealy internodes, very prominent dark purple nodes when young, culm-sheaths sparsely spotted and smaller leaf-blades.

Native to China. It was introduced to France in 1846 and subsequently to England. It grows very well in Europe.

5. Phyllostachys bambusoides *Sieb. & Zucc.*

Phyllostachys mazelii A. & C. Riv. in synon.
P. quiloi A. & C. Riv.
P. reticulata K. Koch non *Bambusa reticulata* Rupr.

English: Giant timber bamboo
Chinese: Gui zhu
Japanese: Madake

5a. var. **bambusoides**

Culms up to 15 m high, 14–16 cm in diameter (in China); internodes green without powder and glabrous when young, the nodes inconspicuous. Culm-sheaths greenish brown, densely spotted or blotched with dark brown, sparsely hirsute; auricles and oral setae small, sometimes lacking on the smaller sheaths. Leaf-blades conspicuously larger than in other species of *Phyllostachys*, up to 15–16 cm long, 2–2.5 cm wide; leaf-sheaths with conspicuous auricles and oral setae.

This bamboo is native to China and is widely distributed from the Yellow River southwards to Guangdong, Guangsi and Yunnan provinces. It is extensively cultivated in Japan, and was introduced from there into France by Admiral Du

Quilio in 1866, thence into Britain circa 1890.

P. bambusoides is the largest and most commercially valuable bamboo in China and Japan after *P. edulis*, the wood quality being very good. It is very hardy and its shoots emerge later than in any other species of *Phyllostachys*, usually towards the end of May. It is a beautiful bamboo with glossy green internodes.

5b. f. subvariegata (*Makino*) *Muroi*

Phyllostachys reticulata K. Koch. f. *subvariegata* Makino

Differs only in the light green stripes on its leaf-blades.

5c. var. castillonii (*Carr.*) *Makino*

Bambusa castillonii Carr.
Phyllostachys castillonii (Carr.) Mitf.
P. bambusoides cv. Castillon

English: Castillon bamboo
Japanese: Kinmei-chiku

It is distinguished from the typical variety of *P. bambusoides* by its bright yellow internodes with green grooves and by its smaller stature.

The variety is of Chinese origin, but is was introduced into Japan at an early date and from there was brought to Europe about 1886.

This attractive bamboo is useful as an ornamental where space is no problem or in a small space if its running rhizomes can be confined.

6. Phyllostachys bissetii *McClure*

English: David Bisset bamboo
Chinese: Bai-jia zhu

Culms of medium size, up to 5–6 m high, about 2 cm in diameter, dark green tinted more or less dark purple, sparsely hirsute below the nodes. Culm-sheaths pale brownish tinged with purple, without spots, glabrous except for the margin, more or less glaucous; blades lanceolate or narrowly triangular, slightly reflexed; auricles usually developed on the midculm sheaths.

This species much resembles *P. heteroclada*, but can be recognized by the conspicuous auricles and narrow blades of the culm-sheaths.

It is native to W. China. It was introduced into the United States in 1941 and into Britain a couple of years ago, growing at a private garden in W. Sussex.

7. Phyllostachys heteroclada *Oliv.*

Phyllostachys cerata McClure
P. congesta Rendle
P. purpurata McClure

English: Fishscale bamboo
Chinese: Shui zhu (Water bamboo)

Culms 3–4 m high, 2–3 cm in diameter; internodes at midculm 30 cm long, loosely mealy and sparsely retrorsely scabrous below nodes when young. Culm-sheaths shorter than internodes, green or dark green with many purple veins, glabrous but densely fringed with minute cilia at the margin; blades triangular or triangular-lanceolate, green, erect; auricles very small with long oral setae. Leaf-blades usually 2 on each branchlet.

The species is native to China and always found along the banks of streams in the mountains. It was first introduced to the United States by Frank N. Meyer in 1907, and was taken from there to Europe.

Not commonly cultivated in Britain.

8. Phyllostachys aureosulcata *McClure*

English: Yellow-groove bamboo, Stoke and Forage bamboo
Chinese: Huang-cao zhu (Yellow-groove bamboo), Yu-ziang-jin zhu (Jade inlaid gold bamboo)

Culms 4–6 m high, 2–4 cm in diameter, sometimes more or less zigzag; internodes 15–20 cm long, green usually with a yellow groove, densely retrorsely scabrous, nodes very prominent. Culm-sheaths brownish green with pale wine and cream streaks, glabrous, with very small spots; blades triangular or triangular-lanceolate, decurrent on both sides into the auricles; auricles and oral setae conspicuous.

This bamboo is native to China, and is cultivated as far north as Beijing. It was introduced from Zhejiang province of China to the United States in 1907, and recently taken from there to the Royal Botanic Gardens, Kew. It is hardy and grows very well, flowering at Kew in 1983.

It is one of the most beautiful garden bamboos and is widely planted in China.

9. Phyllostachys mannii *Gamble*

Phyllostachys assamica Brandis
P. bawa E.G. Camus
P. decora McClure

Chinese: Huang-ku zhu (Yellow bitter bamboo)
Burmese: Mai-pang puk

Culms up to 8–9 m high, 4–6 cm in diameter; internodes 27–42 cm long, bright green when young, nodes scarcely prominent. Culm-sheaths green-yellow with dense purple streaks and small spots, glabrous, tinted purple-red at the margin, broadly truncate or broadly rotund at apex; auricles small on the lower sheaths and falcate on the upper sheaths; ligules purple, white-ciliate at the apex and with purple bristles from the back.

Very similar to *P. aureosulcata* but the culms of the latter have yellow grooves and very prominent nodes, the culm-sheaths are narrower, and the ligule lacks bristles.

P. mannii is native to China and India, with a wide distribution and ecological adaptability. It was introduced from Jiangsu province of China to the United States in 1938, and has recently been introduced to Europe.

The culms are used for woven bamboo articles and are of very good quality. The shoots taste bitter and are not normally utilized for cooking.

10. Phyllostachys meyeri *McClure*

English: Meyer bamboo
Chinese: Zhejiang Dan zhu (Zhejiang lightly sweet bamboo), Dan zhu.

Culms up to 11 m high and 7 cm in diameter, erect and straight; internodes glaucous, glabrous except for minutely hairy sheath-scars. Culm-sheaths yellowish brown, spotted and more or less mottled with dark brown areas, glabrous except at the base, without auricles or oral setae; blades narrowly strap-shaped, very long, arched or pendulous.

P. meyeri is similar to *P. aurea*, but the lower culm-internodes of *P. aurea* are often shortened and swollen, and its culm-sheaths are lighter in colour and only sparsely spotted.

Native to E. China. It was first taken from Zhejiang province to the United States in 1907, and was introduced to Britain a couple of years ago.

P. meyeri is one of the most valuable bamboos in China, especially in Zhejiang, Anhwei and Jiangzu provinces. The culms of this species are of good quality for fishing rods and woven bamboo articles. The shoots are quite good for cooking.

11. **Phyllostachys aurea** A. & C. Riv.

Phyllostachys bambusoides Sieb. & Zucc. var. *aurea* (A. & C. Riv.) Makino
P. formosana Hayata

English: Fishpole bamboo
Chinese: Luo-han zhu (Buddha's face bamboo), Ren-mian zhu (Human face bamboo).
Japanese: Hotei-chiku

Culms 5–6 m high, 2–3 cm in diameter, erect and straight; internodes green, gradually becoming yellowish, usually shortened and swollen in the lower part of the culm, sometimes with the nodes obliquely inclined; sheath-scars minutely pubescent when young. Culm-sheaths brownish green with a few small spots, glabrous except for the base; blades long, strap-shaped, pendulous; auricles and oral setae undeveloped; ligules long-ciliate.

This species is native to China, but has been cultivated in Japan for several centuries. It has been widely introduced, and was brought to Europe in the middle seventies of last century. It first flowered in Britain in 1876, and has since flowered in 1919–1920, 1936–1937 and 1967.

It is an excellent ornamental species. Selected culms with unusual internode patterns are used for fishing rods and walking sticks.

11a. cv. 'Flavescens-inversa'

Distinguished by the green culms with light yellow grooves, and by the leaf-blades often being striped with light yellow.

A plant of garden origin, introduced to Britain a couple of years ago.

12. **Phyllostachys sulphurea** (*Carr.*) A. & C. Riv.

Bambusa sulphurea Carr.
Phyllostachys bambusoides Sieb. & Zucc. var. *sulphurea* (Carr.) Tsuboi
P. castillonii (Carr.) Mitf. var. *holochrysa* Pfitz.
P. reticulata K. Koch var. *holochrysa* (Pfitz.) Nakai
P. reticulata var. *sulphurea* (Carr.) Makino
P. bambusoides cv. Allgold.

English: Sulphur bamboo, Allgold bamboo.
Chinese: Jin zhu (Golden bamboo), Huang pi gong zhu (Yellow skin firm bamboo).

Japanese: Kin-chiku.

12a. var. **sulphurea**

Culms 7–8 m high, 3–4 cm in diameter; internodes golden yellow with a few vertical green stripes, glabrous, only one prominent ridge (sheath-scar) on the lower nodes. Culm-sheaths yellow-green with a few brown spots, glabrous, without auricles and oral setae. Leaf-blades sometimes with yellowish or cream stripes.
Native to E. China, occurring sporadically in forests of *P. sulphurea* var. *viridis*. It was introduced to France in 1865.
This is a very hardy and attractive bamboo, but its cultivation in Britain is limited.

12b. var. **viridis** *R.A. Young*

Phyllostachys faberi Rendle
P. mitis A. & C. Riv. non *Bambusa mitis* Poiret
P. viridis (Young) McClure

English: Green sulphur bamboo.
Chinese: Gang zhu (Firm bamboo).

The variety is distinguished from *P. sulphurea* var. *sulphurea* by its green or bright green culms, and its ultimate size is also considerably larger.
Native to China. Introduced to Europe in 1856.
This is the common wild form of the species and is one of the most important bamboos for commercial production in China. Its culms, which have excellent technical properties, are chiefly used for the handles of farm tools. The shoots taste a little bitter when fresh and it is best to cut them into slices, parboil them for a few minutes, and change the water. They are then quite satisfactory for culinary purposes.

13. **Phyllostachys nuda** *McClure*

Chinese: Shi zhu (Stone bamboo - alluding to the heavy thick-walled culms).

Culms up to 8 m high, 3–4 cm in diameter with thick walls, dark green, copiously and loosely mealy at sheath-fall, glabrous, the nodes very prominent. Culm-sheaths pale purplish brown with numerous purple veins, mealy and mottled, glabrous but somewhat scabrous; blades short, triangular-lanceolate; auricles and oral setae lacking; ligules truncate at the apex.
P. nuda is easily recognized by its thick-walled culms, distinctively prominent nodes and more or less scabrous culm-sheaths.
Native to E. China, forming extensive forests at lower elevations in the mountains. It was introduced from Zhejiang province to the United States in 1908, and to Britain recently. It grows well.
In China the culms of this species are often used for the legs of bamboo furniture. The shoots are quite pleasant for cooking when fresh and are dried for export.

14. **Phyllostachys makinoi** *Hayata*

English: Makino bamboo.
Chinese: Taiwan Gui zhu, Gui zhu

Culms up to 18 m high, 7–8 cm in diameter; internodes light green with copious powder at sheath-fall, glabrous, the nodes marked by two prominent ridges. Culm-sheaths glabrous, with black or smoke-brown spots; auricles and oral setae lacking; ligules bristly with long reddish hairs when fresh. Leaf-sheaths bearing conspicuous auricles and oral setae.

This species is similar to *P. bambusoides* and *P. sulphurea* var. *viridis*. From the former, it is distinguished by its glaucous internodes at sheath-fall, its glabrous culm-sheaths without auricles and its reddish ligular bristles. From the latter, by the two prominent ridges on its nodes and also by the ligular hairs.

Native to China. It is the most important and valuable bamboo in Taiwan, but is rarely cultivated in Britain.

15. Phyllostachys vivax *McClure*

English: Smooth-sheath bamboo.
Chinese: Wu-ke-bu-ji zhu, Wu-bu-ji zhu (Dark sheath bamboo).

Culms 10–15 m high, 4–8 cm in diameter, arched at the top; internodes green, more or less mealy, glabrous. Culm-sheaths yellowish brown, densely spotted and mottled with black-brown areas, glabrous, without auricles or oral setae; blades strongly wrinkled, reflexed; ligules conspicuously decurrent along both sides.

P. vivax is likely to be confused with *P. bambusoides*. It may be distinguished by its glaucous internodes at sheath-fall, glabrous culm sheaths, entire lack of auricles and oral setae, strongly decurrent ligules and wrinkled blades. Its new shoot growth is much earlier than that of *P. bambusoides*.

It is native to E. China, and commonly cultivated in the lowlands around villages. It was introduced from Zhejiang province to the United States in 1908, and brought to Britain several years ago, growing in a private garden in W. Sussex.

This bamboo is vigorous, reaching maturity quickly. It is also a very important edible bamboo whose shoots are entirely free from bitter flavour, even when fresh. It has therefore been extensively planted for commercial production in E. China. The culms are thin-walled and of limited industrial value.

16. Phyllostachys angusta *McClure*

Chinese: Huang-gu zhu (vernacular name of Zhejiang province).

Culms 8 m high, 2–4 cm in diameter, straight, bearing a narrow pyramidal crown of foliage; midculm-internodes about 26 cm long, green to grey-green, glabrous. Culm-sheaths yellowish white with a few purplish veins and small sparse spots, glabrous; blades strap-shaped, green with yellowish margin, not wrinkled, reflexed; auricles and oral setae undeveloped; ligules yellowish green, very prominent, laciniate, fringed with long white hairs.

This species is characterized by straight culms with narrow crown, pale culm-sheaths and long hairy ligules.

P. angusta is native to E. China and grows in bamboo forest mixed with other species. It was introduced from Zhejiang province to the United States by Frank N. Meyer in 1907, whose field notes mistakenly called it 'Sah Chu' (Stone bamboo), a name properly ascribed to *P. nuda*. It was introduced to Britain recently, growing in a private garden in W. Sussex.

The culms of this species are mostly used for the manufacture of fine bamboo articles and are of better quality than those of *P. meyeri*, *P. edulis* and *P. glauca*.

17. Phyllostachys rubromarginata *McClure*

Chinese: Hong-bian zhu (Red-margined sheath bamboo), Nu-er zhu (Girl's bamboo).

Culms 5–7 m high, 2 cm in diameter, arched at the top; internodes 25-32 cm long, green with thin powder, nodes scarcely prominent. Culm-sheaths shorter than internodes, green with red margins, glabrous, unspotted, without auricles or oral setae; blades strap-shaped to lanceolate; ligules purple, long-ciliate at apex. Leaf-blades 7.5–15 cm long, 1.3–2.1 cm wide; leaf-sheaths with long reddish oral setae, auricles undeveloped.

This species is characterized by long slender internodes, unspotted culm-sheaths with red margins and by the long-ciliate ligules of the culm-sheaths.

Native to China. It was introduced from S. China to the United States by McClure and brought to Britain a couple of years ago, growing in a private garden in W. Sussex.

18. Phyllostachys propinqua *McClure*

Chinese: Jiao-ku dan zhu (Withered sheath bamboo)

Culms 10 m high, 5 cm in diameter; internodes bright green with thin powder, glabrous. Culm-sheaths yellowish brown with powder, usually withered at the margins, entirely glabrous, with small spots; blades narrowly strap-shaped, reflexed; auricles and oral setae lacking; ligules more or less convex.

P. propinqua is most apt to be confused with *P. meyeri*, but lacks the narrow line of white hairs found on the sheath-scars and at the base of the culm-sheaths in the latter.

Native to China, and very widely distributed from S. Honan southwards to Guizhou and Guangxi provinces. It was introduced from Guangxi province to the United States in 1928 and recently to Britain where it is growing in a private garden in W. Sussex.

19. Phyllostachys glauca *McClure*

Chinese: Dan zhu (Lightly sweet bamboo).

Culms 10–12 m high, 2–5 cm in diameter, arched at the top; internodes up to 30–40 cm long, glaucous with dense powder, glabrous, the nodes slightly prominent. Culm-sheaths green with densely purple veins, purple-brown spotted or mottled (sparsely so or absent on the smaller sheaths); blades short; auricles and oral setae lacking; ligules purple or purple-brown, 1–3(–4) mm long, truncate at the apex, with short cilia. Leaf-sheaths with purple ligules and deciduous auricles and oral setae.

The species is very similar to *P. flexuosa*, but culms of the latter are often somewhat zigzag, and the ligules of its culm-sheaths are dark maroon and longer than those of *P. glauca*.

Native to N. China from the Yellow to the Yangtze River, and very hardy. It was introduced from Nanjing to the United States by McClure in 1926 but not given a name until 1956. It reached Britain recently, growing in a private garden in W. Sussex.

It is one of the most important and valuable bamboos for commercial production in N. China. The culms are used for fishing rods, woven bamboo baskets, bamboo mats and other articles. The shoots taste quite good when cooked.

20. Phyllostachys flexuosa *A. & C. Riv.*

Bambusa flexuosa Carr. non Munro

Chinese: Tian zhu (Sweet bamboo).

Culms 5–6 m high, 2–3 cm in diameter; midculm internodes 25–30 cm long, sometimes zigzag, more or less glaucous, glabrous, the nodes slightly prominent. Culm-sheaths greenish brown with pale purple veins, spotted; blades long and narrow, strongly arched; auricles and oral setae undeveloped; ligules long with long bristles at the apex.

Native to N. China. Introduced to France in 1864, and now widely cultivated in Europe. It is very hardy and grows quite well in Britain.

The culms are used for fishing rods and woven bamboo articles. The shoots are edible.

2. Shibataea *Nakai*

Small shrubs; rhizomes monopodial with spaced culms. Culm-internodes conspicuously flattened on one side, the nodes swollen and very prominent; branch-complement 3–5, the branches very short, with 1–2 internodes bearing 1 or 2 leaf-blades, the terminal leaf with an obsolete sheath; culm-sheaths thinly chartaceous. Inflorescence comprising 1-spikelet racemes subtended by sheathing bracts. Spikelets without pedicels, 2-flowered; glumes 2–3; stamens 3; stigmas 3 on a long style.

This is a peculiar genus, easily recognized by its small stature with conspicuously flattened culms and abbreviated primary branches.

5 species. China and Japan.

Shibataea kumasaca *(Steud.) Nakai*

Bambusa kumasaca Steud.
Phyllostachys kumasaca (Steud.) Munro
Shibataea ruscifolia Makino

English: Ruscus-leaved bamboo
Chinese: Riben Emao zhu (Japanese goose feather bamboo)
Japanese: Okame-zasa, Gomai-zasa, Bungo-zasa

Culms 0.5–1.5 m high, 3–5 mm in diameter, remarkably zigzag and flattened on one side; branches 3–5 per node, very short, usually bearing a single leaf, rarely 2. Culm-sheaths thin, glabrous. Leaf-blades oblong-lanceolate or ovate-lanceolate, 5–11 cm long, 2–2.5 cm wide, pubescent beneath, without oral setae, the sheaths of terminal leaves obsolete.

Native to Japan. It was introduced to Britain in 1861, and is widely planted in many parts of Europe. It requires plenty of moisture. Flowering in Britain was recorded in 1964.

The dwarf habit and squat broad leaf-blades make it a very attractive ornamental bamboo.

3. Chimonobambusa *Makino*

Small trees and shrubs; rhizomes monopodial with spaced culms. Culm-internodes grooved on one side, the nodes swollen, very prominent, and often with root-thorns; branch-complement 3, subequal or with a few smaller secondary branches. Culm-sheaths usually papery, tardily deciduous or persistent; blades reduced, very small and inconspicuous. Inflorescence comprising 1–3 single-spikelet racemes subtended by small bracts. Spikelets many-flowered, sessile; stamens 3; stigmas 2 on a short style.

The genus is characterized by its monopodial rhizomes, swollen and usually thorny culm-nodes, and by the culm-sheaths with very small blades.

About 10 species in China, Japan, India, Burma and Viet-Nam. Two species have been introduced to Britain.

Key to species of *Chimonobambusa*

Small tree up to 4–7 m high; culms quadrangular, with thorns on lower nodes; internodes rough, retrorsely scabrid when young; culm-sheaths shorter than internodes, deciduous, subglabrous **1. C. quadrangularis**
Shrub 2–3 m high; culms cylindrical, usually without thorns on nodes; internodes smooth, glabrous; culm-sheaths longer than internodes, persistent, pubescent at base **2. C. marmorea**

1. Chimonobambusa quadrangularis *(Fenzi) Makino*

Bambusa quadrangularis Fenzi
Arundinaria quadrangularis (Fenzi) Makino
Phyllostachys quadrangularis (Fenzi) Rendle

English: Square bamboo, Square-stemmed bamboo
Chinese: Fang zhu (Square bamboo)
Japanese: Shiho-chiku, Shikaku-dake

Culms 5–8 m high, 2–4 cm in diameter, quadrangular in cross-section with root-thorns on the lower nodes; internodes papillose and rough. Culm-sheaths deciduous, shorter than internodes, straw-coloured with purple or dark purple blotches, conspicuously tessellate, without auricles or oral setae; blades very small, only 1–3 mm long. Leaf-blades narrowly lanceolate, thinly papery, conspicuously tessellate, the sheaths with well developed oral setae.

Easily recognized by its quadrangular culms, these rough to the touch and with thorny nodes.

Native to China. It was introduced to Europe in the last century, but its distribution in Britain is limited to milder regions.

It is a very peculiar bamboo and suits all types of gardens as an ornamental.

2. Chimonobambusa marmorea *(Mitf.) Makino*

Bambusa marmorea Mitf.
Arundinaria marmorea (Mitf.) Makino

English: Marble bamboo
Chinese: Han zhu (Cold bamboo)
Japanese: Kan-chiku

Culms 2–3 m high, 1–1.5 cm in diameter with very thick walls; internodes green at first becoming dull purple at maturity, glabrous except for the sheath-scars, cylindrical, without nodal thorns. Culm-sheaths longer than internodes, persistent, membranous, purplish with purple-brown spots, pubescent at base;

blades none or very small, 1 mm long. Leaf-blades small, the sheaths with soft oral setae.

This is very peculiar species and easily recognized.

Native to Japan. It was introduced to France by M Marliac in 1889 and brought to Britain shortly afterwards. It first flowered in Britain in 1909, and subsequently at Kew Gardens in 1930–1931 and from 1954 to the present day.

C. marmorea is a hardy and attractive bamboo, with an unusual marble-like colouring on the new shoots and culm-sheaths.

4. Sinobambusa *Nakai*

Small trees; rhizomes monopodial with spaced culms. Culm-internodes elongated, more or less grooved on one side, the nodes usually more or less prominent; branch-complement usually 3 at each node, subequal. Culm-sheaths deciduous, abscissing completely, usually with auricles and oral setae; blades strongly developed. Inflorescence with short branches bearing a few spikelets loosely grouped about the nodes. Spikelets very long, linear, many-flowered, the lateral spikelets with a small bract at the base; glumes 2 to several; stamens 3; stigmas 3.

This genus much resembles *Semiarundinaria*, but the latter has incompletely abscissing culm-sheaths, and conspicuous bracts subtending racemes of 1–2 spikelets.

About 10 species in China and Viet-Nam. One species has been introduced to Britain.

Sinobambusa tootsik (*Sieb.*) *Makino*

Bambusa tootsik Sieb.
Arundinaria tootsik (Sieb.) Makino

Chinese: Tang zhu
Japanese: To-chiku

Culm 5–10 m high, 2–4 cm in diameter; internodes remarkably elongated, up to 60–80 cm long, glabrous, the nodes prominent and densely pilose with long brown hairs at sheath-scars. Culm-sheaths thick, coriaceous, without spots, sparsely pilose with dark brown hairs becoming densely so at base; blades bright green, lanceolate; auricles falcate and fringed with oral setae. Leaf-blades medium size; leaf-sheaths with auricles and remarkably long radiate oral setae.

Native to China, and introduced to Japan during the Tang dynasty. It has been grown in the Temperate House at Kew Gardens since 1920, but its distribution in Britain is very limited.

5. Semiarundinaria *Nakai*

Small trees or shrubs; rhizomes monopodial with spaced culms. Culm-internodes more or less grooved on one side; branch-complement commonly 3. Culm-sheaths deciduous, abscissing incompletely; blades well developed; auricles absent or rudimentary. Inflorescence comprising fascicled racemes of 1–3 spikelets subtended by spathaceous bracts. Spikelets several-flowered; glumes 0–3; stamens 3; stigmas 3.

This genus is similar to *Sinobambusa*, but the latter has completely abscissing culm-sheaths and an inflorescence of only a few spikelets subtended by small bracts.

6 species in Japan and China. Two species have been introduced to Britain.

Key to species of *Semiarundinaria*

1 Branches 3 or more on each node; culm-sheaths glabrous except at the base, without auricles or oral setae; leaf-blades 1.5–2.5 cm wide **1. S. fastuosa**
1 Branches 1 or 2 on each node; culm-sheaths pubescent, with auricles and oral setae; leaf-blades 3.5–5 cm wide: 2
  2 Leaf-blades uniformly green **2. S. tranquillans**
  2 Leaf-blades green with yellow stripes **2a. S. tranquillans** cv. Shiroshima

1. **Semiarundinaria fastuosa** (*Mitf.*) *Nakai*

Bambusa fastuosa Mitf.
Arundinaria narihira Makino
Phyllostachys fastuosa (Mitf.) Makino

English: Narihira bamboo
Chinese: Ye-ping zhu
Japanese: Narihira-dake

Culms 5–8 m high, 2–4 cm in diameter, thin-walled, upright, green at first, more or less purplish later, the nodes slightly prominent; branch-complement 3, sometimes more but then 3 of them dominant. Culm-sheaths coriaceous, green with purplish streaks, glabrous except for the puberulous base; auricles and oral setae absent or rudimentary. Leaf-blades narrowly lanceolate, 18–24 cm long, 1.5–2.5 cm wide, with straight oral setae and no auricles.

Native to Japan. It was introduced to France by M Marliac in 1892, and brought to Britain in 1895 where it flowered in 1935–1936, 1957 and 1965–1970.

This species is an exceptionally stately and handsome bamboo. It is very hardy and withstands temperatures of −20°C.

2. **Semiarundinaria tranquillans** *Koidz.*

Phyllostachys tranquillans (Koidz.) Muroi

Culms 1.5–2 m high, 5 mm in diameter; internodes glabrous, about 20 cm long; branch-complement 1 or 2. Culm-sheaths shorter than internodes, subcoriaceous, white-pubescent; blades narrowly lanceolate, erect; auricles developed but small, with oral setae. Leaf-blades large, oblong-lanceolate, 15–22 cm long, 3–4.5 cm wide, glabrous; leaf-sheaths more or less pubescent, with conspicuous auricles and radiate oral setae; ligules conspicuously long, ciliate at apex.

Native to Japan. It was introduced to Kew Gardens several years ago, and is growing well.

2a. cv. 'Shiroshima'

Hibanobambusa tranquillans (Koidz.) Maruyama & Okamura f. *shiroshima* Okamura.

Distinguished from the typical form of the species by its yellow-striped leaf-blades.

Native to Japan. It has been cultivated in the Kew bamboo garden. It resembles *Arundinaria argenteostriata*, but the latter has the culm-sheaths glabrous except at the base, and smaller leaf-blades.

It is an attractive foliage bamboo.

6. Arundinaria *Michx.*

Pleioblastus Nakai
Nipponocalamus Nakai

Small trees and shrubs; rhizomes monopodial. Culm-internodes terete, without a groove; branch-complement usually 3–7, thinner than culm. Culm-sheaths tardily deciduous or persistent. Inflorescence usually racemose, sometimes reduced to a spikelet. Spikelets several to many-flowered with pedicels; glumes 2; stamens 3; stigmas 3, rarely 2.

The genus *Arundinaria* has been interpreted by different authors in various ways creating a good deal of nomenclatural confusion. It is now limited to species that conform to the generic description given above.

About 50 species. Native to North America, Japan, China and South Asia, the majority of the species in Japan and China. Ten species have been introduced to Britain. They all grow very well.

Key to species of *Arundinaria*

1 Culms 3–5 m high; branches numerous at each node: 2
  2 Young culm-sheaths densely pilose, becoming glabrous with age; leaf-blades linear, usually 0.8–1.2 cm wide **1. A. graminea**
  2 Young culm-sheaths glabrous, or sometimes puberulous at the base; leaf-blades linear-lanceolate, 1–2(–2.5) cm wide: 3
    3 Leaf-blades uniformly green **2a. A. simonii f. simonii**
    3 Leaf-blades green with white stripes **2b. A. simonii f. variegata**
1 Culms 1–2 m high or less; branches usually 2–3 at each node: 4
  4 Leaf-blades variegated: 5
    5 Leaf-blades glabrous beneath, green with white or yellowish stripes, 2–3 cm wide **3. A. argenteostriata**
    5 Leaf-blades pubescent beneath: 6
      6 Leaf-blades light or yellowish green with dark green stipes, 1.5–2.5 cm wide **4. A. auricoma**
      6 Leaf-blades green with white or yellow stripes, 1–1.5 cm wide **5. A. fortunei**
  4 Leaf-blades uniformly green: 7
    7 Culms 20–40 cm high; leaf blades very small, 3–7 cm long, 3–8 mm wide, quite glabrous **6. A. pygmaea var. disticha**
    7 Culms 1.5–2 m high; leaf-blades over 10 cm long, 1 cm wide: 8
      8 Culm-sheaths pubescent **7. A. gigantea var. tecta**
      8 Culm-sheaths glabrous, or pubescent only at base: 9
        9 Culm-sheaths pubescent at the base; leaf-blades lanceolate: 10
          10 Leaf-blades pubescent on both sides **8. A. pumila**
          10 Leaf-blades glabrous on both sides **9. A. humilis**
        9 Culm-sheaths glabrous; leaf-blades narrowly lanceolate, glabrous: 11
          11 Leaf-blades more than 1 cm wide **10a. A. chino f. chino**
          11 Leaf-blades 5–8 mm wide **10b. A. chino f. angustifolia**

1. **Arundinaria graminea** (*Bean*) *Makino*

Arundinaria hindsii Munro var. *graminea* Bean
Pleioblastus gramineus (Bean) Nakai
Thamnocalamus hindsii (Munro) E.G. Camus var. *graminea* (Bean) E.G. Camus

Chinese: Da-ming zhu
Japanese: Taimon-chiku, Tsushi-chiku

Culms 3–5 m high, 1–2 cm in diameter, from a short rhizome; internodes 20–25 cm long, green or yellow-green, glabrous; branch-complement many. Culm-sheaths green, densely pilose at first, glabrous later, without auricles or oral setae; blades small, linear, 2–3 cm long, 1–2 mm wide. Leaf-blades very narrow, linear or lanceolate-linear, 15–27 cm long, 0.8–1.2(–2) cm wide, caudately acuminate at the apex, cuneate at the base, glabrous, glaucous beneath; leaf-sheaths usually without auricles or oral setae, sometimes with a few erect oral setae; ligules conspicuously long.

The species is characterized by its short rhizomes and its very slender leaf-blades.

Native to the Ryukyu Islands of Japan. It is believed to have been introduced to Britain to by Messrs Veitch in 1877, and is known to have flowered several times since 1948.

It is a good species of *Arundinaria* for gardens, as its short rhizomes are easy to control.

'*A. hindsii*' is a broader leaved relative of *A. graminea* sometimes grown in Britain, though of no great horticultural merit. It is not the Chinese *A. hindsii* Munro, but appears to be a Japanese species whose correct name is still uncertain.

2. **Arundinaria simonii** (*Carr.*) *A. & C. Riv.*

Bambusa simonii Carr.
Nipponocalamus simonii (Carr.) Nakai
Pleioblastus simonii (Carr.) Nakai

English: Simon bamboo
Japanese: Me-dake, Kawa-take

2a. f. **simonii**

Culms 4–5(–8) m high, 2–3 cm in diameter, erect; internodes 30–40 cm long, dark green, glabrous, the nodes glabrous or sometimes densely puberulous, not prominent; branch-complement many. Culm-sheaths green, glabrous, sometimes densely puberulous at the base, conspicuously concave at the apex, without auricles or oral setae; ligules truncate or concave at the apex; blades linear to linear-lanceolate, green. Leaf-blades lanceolate to linear-lanceolate, 15–25 cm long, 1–2(–2.5) cm wide, long-acuminate, glabrous, vivid green above, green on one side and greyish on the other beneath; leaf-sheaths with a few straight white oral setae.

This is a very common *Arundinaria* in Japan. It may be distinguished from other species by the following combination of characters: tall olive-green culms, long internodes, many branches at each node, glabrous culm-sheaths with concave apex, and two quite distinct shades of colour on the underside of the leaf-blades.

It has been introduced to many countries for gardens, and is believed to have been introduced to France in 1862 by M Simon. It was brought over to Britain shortly afterwards, and has flowered several times, the first being in 1903–1905 and the latest in 1967.

This bamboo is very hardy and vigorous, and is ideal for garden use though sometimes frost damage to the late autumn growth gives it a shabby appearance in winter.

2b. f. variegata (*Hook. f.*) *Rehd.*

Arundinaria simonii var. *variegata* Hook.f.
A. simonii var. *albo-striata* Bean
Nipponocalamus simonii (Carr.) Nakai var. *variegatus* (Hook.f.) Nakai
Pleioblastus simonii (Carr.) Nakai var. *variegatus* (Hook.f.) Nakai
P. simonii f. *variegatus* (Hook.f.) Nakai

Distinguished from the typical form by longitudinally white or yellowish stripes on the leaf-blades.

3. Arundinaria argenteostriata (*Regel*) *Vilmorin*

Bambusa argenteostriata Regel
Arundinaria chino (Franch. & Sav.) Makino var. *argenteostriata* (Regel) Makino
Nipponocalamus argenteostriatus (Regel) Nakai
Pleioblastus argenteostriatus (Regel) Nakai
Sasa argenteostriata (Regel) E.G. Camus

Japanese: Okina-dake

Culms 0.5–1.5 m high, 2–3 mm in diameter; internodes 15–25 cm long, green, glabrous, the nodes sometimes puberulous; branch-complement 1–3. Culm-sheaths shorter than internodes, glabrous except for the pubescent base and long-ciliate margins, without auricles or oral setae. Leaf-blades lanceolate, 14–20 cm long, 2–2.5 cm wide, glabrous on both sides, with a few white or yellowish stripes; leaf-sheaths glabrous except for the margins and the long white oral setae.
Native to Japan. It has been introduced to Britain but is not generally cultivated.

4. Arundinaria auricoma *Mitf.*

A. variabilis Vilm. var. *viridi-striata* Makino
A. variegata (Miq.) Makino var. *viridi-striata* (Makino) Makino
A. viridi-striata (Makino) Nakai
Bambusa viridi-striata Regel (1867)
B. viridi-striata André (1872) non Regel
Pleioblastus viridi-striatus (Makino) Makino
Sasa auricoma (Mitf.) E.G. Camus

Japanese: Kamuro-zasa

Culms 1.5 m high, 3–4 mm in diameter, simple or branching; internodes 15–17 cm long, green, shortly pubescent; branch-complement usually 1 on each node during first year, one or two more in the second year. Culm-sheaths shorter than internodes, with inconspicuous hairs, without auricles or oral setae. Leaf-blades oblong-lanceolate, 15–20 cm long, 2–3 cm wide, acuminate at the apex, rounded at the base, light green or yellowish with dark green stripes, sparsely pubescent above, densely pubescent beneath; leaf-sheaths with minute hairs; oral setae sparse or lacking.

The species is easily recognized by its dwarf habit and large pubescent leaf-blades with dark green stripes.

Native to Japan, and widely cultivated there as a decorative plant but its natural habitat unknown. It was introduced into Europe in the 1870s, flowering in Britain in 1898 and again in 1935. The plant was first named *Bambusa viridistriata* by two botanists working independently in Russia and France, using different plants from the same batch of living material collected in Japan. Unfortunately subsequent combinations were all based on the later, illegitimate, homonym allowing *A. auricoma* to take priority.

This is an attractive dwarf bamboo for growing in the ornamental border or foliage garden where it can be allowed to display its beautiful foliage to full advantage.

5. Arundinaria fortunei *(Van Houtte) Nakai*

Bambusa fortunei Van Houtte
Arundinaria variabilis Vilm. var. *variegata* (Miq.) Makino
A. variegata (Miq.) Makino
Bambusa variegata Miq.
Nipponocalamus fortunei (Van Houtte) Nakai
Pleioblastus fortunei (Van Houtte) Nakai
P. variegatus (Miq.) Makino
Sasa fortunei (Van Houtte) Fiori
Sasa variegata (Miq.) E.G. Camus

English: Dwarf white-stripe bamboo
Chinese: Fei-bai zhu
Japanese: Chigo-zasa, Shima-dake

Culms 0.3–1 m high, 1–2 mm in diameter; internodes 12–17 cm long, light green, glabrous; branch-complement 1–2. Culm-sheaths shorter than internodes, glabrous, without auricles or oral setae. Leaf-blades lanceolate, 10–15(–20) cm long, 1–1.5 cm wide, green with yellowish or white stripes, sparsely pubescent or subglabrous above, densely pubescent beneath; leaf-sheaths glabrous except the margins; oral setae sparse and straight or lacking.

This species is similar to *A. auricoma* and *A. argenteostriata* in possessing variegated leaf-blades. From *A. auricoma* it is distinguished by its narrower leaf-blades with yellowish or white stripes, and from *A. argenteostriata* by its pubescent leaf-blades.

Native to Japan, where it is widely cultivated. It was first introduced to Belgium by Van Houtte of Ghent before 1863 and brought to Britain about 1876. The only record of flowering is from Argentina (Suzuki in J. Jap. Bot. 54: 183–184, 1979).

6. Arundinaria pygmaea (Miq.) Aschers. & Graeb. var. **disticha** *(Mitf.) Chao & Renv.*

Bambusa disticha Mitf.
Arundinaria variabilis Vilm. f. *glabra* Makino
Nipponocalamus argenteostriatus (Regel) Nakai var. *distichus* (Mitf.) Ohwi
Pleioblastus distichus (Mitf.) Nakai
P. pygmaeus (Miq.) Nakai var. *distichus* (Mitf.) Nakai
Sasa disticha (Mitf.) E.G. Camus

English: Dwarf fern-leaf bamboo
Chinese: Cui zhu (Emerald green bamboo)

6. Arundinaria

Japanese: Oroshima-chiku

Culms 20–40(–100) cm high, 1–2 mm in diameter; internodes 10–15 cm long, glabrous; branch-complement 1–2, the branches rather long and bearing many leaves arranged in two opposite rows. Culm-sheaths glabrous except the margins. Leaf-blades very small, 3–7 cm long, 3–8(–10) mm wide, usually glabrous, light green above, slightly glaucous beneath; leaf-sheaths ciliate at the margins, with a few oral setae.

This variety differs from the typical form of *A. pygmaea* in its glabrous leaves and culm-sheaths. It is distinguished from the other *Arundinaria* species by its smaller habit and its small distichous leaves.

It is only known in cultivation. In Japan it is widely used as a decorative bamboo. It was introduced to Britain by Messrs Veitch about 1870, and distributed under the erroneous name '*Bambusa nana*'. It flowered in Britain in 1967–1970.

This is an attractive low-growing bamboo, particularly suitable for growing in rockeries. It also makes a very good tub plant.

7. **Arundinaria gigantea** (*Walt.*) *Muhl.* subsp. **tecta** (*Walt.*) *McClure*

Arundinaria tecta (Walt.) Muhl.
A. macrosperma Michx. var. *suffruticosa* Munro
A. macrosperma var. *tecta* (Walt.) Wood

English: Switch cane, Small cane

Culms 2 m high, 1–1.5 cm in diameter; internodes 14–18 cm long, green, glabrous; branch-complement usually 2–3. Culm-sheaths shorter than internodes, densely setose with deciduous hairs; auricles and oral setae strongly developed. Leaf-blades lanceolate, 11–23 cm long, 1.3–2.5 cm wide, glaucous and pubescent beneath; leaf-sheaths with conspicuous auricles and oral setae.

Native to the SE. United States. It is distinguished from typical *A. gigantea*, which can be up to 8 m high, by its small stature.

The date of introduction into Europe is uncertain, but it reached Britain before 1881, flowering in 1948 at Kew.

This is a second-rate ornamental species for gardens, looking particularly sad during the winter months.

8. **Arundinaria pumila** *Mitf.*

Arundinaria variabilis Vilm. var. *pumila* (Mitf.) H. de Lehaie
Nipponocalamus pumilus (Mitf.) Nakai
Pleioblastus argenteostriatus (Regel) Nakai f. *pumilus* (Mitf.) Muroi
P. chino (Franch. & Sav.) Makino f. *pumilus* (Mitf.) Suzuki
P. pumilus (Mitf.) Nakai
Sasa pumila (Mitf.) E.G. Camus

Japanese: Sadore-yoshi

Culms 1–2 m high, 1–2 mm in diameter, slender; internodes 15–30 cm long, green, glabrous, the sheath scars densely yellow-grey pubescent; branch-complement 1–3. Culm-sheaths green, glabrous except for the base which is densely pubescent with long yellowish grey hairs, without auricles or oral setae; blades small, reflexed. Leaf-blades lanceolate, 12–23 cm long, 1–2.5 cm wide, shortly acuminate at the apex, rounded at the base, sparsely pubescent above, densely pubescent beneath; leaf-sheaths with sparse oral setae or none.

This bamboo has been treated as a form of *Arundinaria chino* but is quite unlike that species. It much resembles *A. humilis* but has pubescent leaf-blades.

Native to Japan. It was introduced to Britain in the latter part of last century, flowering at Wakehurst Place, Sussex in 1968.

It is a pretty and extremely hardy dwarf bamboo which grows very well in Britain.

9. Arundinaria humilis *Mitf.*

Arundinaria nagashima Mitf.
Nipponocalamus humilis (Mitf.) Nakai
Pleioblastus humilis (Mitf.) Nakai
Sasa humilis (Mitf.) E.G. Camus

Japanese: Tovooka-zasa

Culms 1.5–2 m high, 3–7 mm in diameter; internodes 12–22 cm long, green or purplish when young, glabrous, the sheath-scars densely yellow-grey pubescent; branch-complement 1–3. Culm-sheaths densely pubescent with long yellowish grey hairs at the base and conspicuously ciliate on the margins, otherwise glabrous, with small auricles and a few oral setae. Leaf-blades oblong-lanceolate or lanceolate, 10–20 cm long, 1.3–2.5 cm wide, acuminate at the apex, rounded at the base, entirely glabrous on both sides; leaf-sheaths with long straight oral setae.

A. humilis resembles *A. pumila* in habit, leaf-blade shape and pubescent sheath-scars, but differs in its entirely glabrous leaf-blades. Perhaps they are no more than two forms of the same species.

Native to Japan. It was introduced to Britain in the latter part of last century, and flowered in 1964–1967.

10. Arundinaria chino *(Franch. & Sav.) Makino*

Bambusa chino Franch. & Sav.
Arundinaria simonii (Carr.) A. & C. Riv. var. *chino* (Franch. & Sav.) Makino
Nipponocalamus chino (Franch. & Sav.) Nakai
Pleioblastus chino (Franch. & Sav.) Nakai

Japanese: Azuma-nezasa

10a. f. chino

Culms 2 m high, 1–1.5 cm in diameter; internodes 15–20 cm long, glabrous; branch-complement several. Culm-sheaths glabrous, without auricles or oral setae. Leaf-blades narrowly lanceolate, 12–20 cm long, 1–2 cm wide, long-acuminate at the apex, broadly cuneate at the base; leaf-sheaths glabrous with long oral setae.

A. chino is very similar to *A. simonii*, but the latter has taller culms and two quite distinct shades of colour on the underside of the leaf-blade.

Native to Japan, and introduced to Britain in 1875. It is very hardy, but not a good-looking bamboo especially during the winter months.

10b. f. angustifolia *(Mitf.) Chao & Renv.*

Bambusa angustifolia Mitf.
Pleioblastus chino (Franch. & Sav.) Nakai f. *angustifolius* (Mitf.) Muroi & Okamura

Distinguished from the typical form by its very narrow leaves, 6–8 mm wide. Native to Japan. It flowered at Kew Gardens in 1935.

7. Pseudosasa *Nakai*

Yadakeya Makino

Shrubs; rhizomes monopodial with spaced culms. Culm-internodes terete without a groove, the nodes not prominent; branch-complement single at midculm, subequal to culm in diameter. Culm-sheaths persistent, coriaceous. Inflorescence paniculate. Spikelets several-flowered with pedicels; glumes 2; stamens 3 (rarely 4); stigmas 3 on a short style.

Pseudosasa is very similar to *Sasa* and *Indocalamus* in its rhizome, inflorescence and branch-complement. It differs from *Sasa* in having 3 stamens (rarely 4) and from *Indocalamus* in having 3 stigmas. Moreover, the commonly grown species, *Pseudosasa japonica*, is taller than either of those genera.

4 species. Japan, Korea and Taiwan. Only one species has been introduced to Britain.

Pseudosasa japonica *(Steud.) Nakai*

Arundinaria japonica Steud.
Sasa japonica (Sieb. & Zucc.) Makino
Yadakeya japonica (Sieb. & Zucc.) Makino

English: Arrow bamboo
Chinese: Riben shi zhu (Japanese arrow bamboo)
Japanese: Ya-dake

Culms 3–5 m high, 1–2 cm in diameter, branching above midculm; internodes long, olive-green, glabrous; branch-complement 1. Culm-sheaths persistent, green, as long as internodes or longer, densely pilose with antrorse hairs, without auricles or oral setae. Leaf-blades 4–7 on each branch, lanceolate, 20–35 cm long, 2.5–3.5 cm wide, glabrous; leaf-sheaths usually without oral setae; ligules conspicuously long.

Native to Japan and South Korea. Introduced to France by von Siebold in 1850, and thence to England a few years later. It has been widely cultivated in Britain, flowering first in 1872–1874 and several times since then. It has been in continuous flower for the last 10 years.

P. japonica has attractive foliage and is a valuable subject for decorative planting, especially in rugged places.

8. Sasa *Makino & Shibata*

Neosasamorpha Tatewaki
Nipponobambusa Muroi
Sasaella Makino
Sasamorpha Nakai

Shrubs; rhizomes monopodial or with some sympodial branching interspersed among the runners. Culms erect or ascending, single or branched; internodes terete without a groove, the nodes prominent or not; branch-complement usually 1, subequal to culm in diameter, exceptionally 3 at upper nodes of culm. Culm-sheaths persistent. Leaf-blades large, tessellate. Inflorescence paniculate. Spikelets several-flowered with pedicels; glumes 2; stamens 6; stigmas 3.

About 50 species. Mainly Japan, extending to Korea and China. Eight species have been introduced to Britain.

The genus has attracted an unusual amount of attention from taxonomists. Over 400 taxa have been described, but most of them are based on immaterial vegetative characters and can be reduced to synonymy.

Key to species of *Sasa*

1 Leaf-blades glabrous: 2
  2 Culm-sheaths glabrous: 3
    3 Culms 2–3 m high; leaf-blades very large, 18–30 cm long, 5–8 cm wide: 4
     4 Culm-internodes green   **1a. S. palmata** f. **palmata**
     4 Culm-internodes mottled with purplish brown blotches
                         1b. S. palmata f. **nebulosa**
    3 Culms 2 m high; leaf-blades smaller, 15–20 cm long, (1.5–)2–2.5 cm wide:
                                       5
     5 Leaf-blades uniformly green   **2. S. masamuneana**
     5 Leaf-blades green with yellowish stripes
                      2a. S. masamuneana cv. Albostriata
  2 Culm-sheaths more or less pubescent when young; culms 1–2 m high: 6
    6 Leaf-sheaths without oral setae; leaf-blades 2–2.5 cm wide   **3. S. borealis**
    6 Leaf-sheaths with oral setae; leaf-blades 3.5–5.5 cm wide: 7
     7 Leaf-blades oblong, 10–20 cm long, 4.5–5.5 cm wide, usually with withered
      margins   **4. S. veitchii**
     7 Leaf-blades oblong-lanceolate, 15–26 cm long, 3.5–4.5 cm wide, without
      withered margins   **5. S. tsuboiana**
1 Leaf-blades pubescent or puberulous beneath: 8
  8 Culm-sheaths sparsely pubescent; leaf-sheaths with falcate auricles and long
    purple oral setae   **6. S. takizawana**
  8 Culm-sheaths glabrous; leaf-sheaths with inconspicuous auricles and few
    oral setae: 9
    9 Culm- and leaf-sheaths not mealy; leaf-blades 1.5–2(–2.5) cm wide
                      7. S. ramosa
    9 Culm- and leaf-sheaths conspicuously mealy; leaf-blades 3–4 cm wide
                      8 S. hayatae

1. Sasa palmata (*Burb.*) *E.G. Camus*

Bambusa palmata Burb.
Arundinaria palmata (Burb.) Bean

Japanese: Chimaki-zasa

1a. f. **palmata**

Culms 2–3 m high, 7–10 mm in diameter, often curved; internodes 14–20 cm long, green, glaucous when young. Culm-sheaths shorter than internodes, glabrous, without auricles or oral setae; blades small, lanceolate. Leaf-blades very large, oblong, 18–30 cm long, 5–8 cm wide, thick, coriaceous, glabrous, with 9–14 pairs of secondary veins and yellow midrib; leaf-sheaths glabrous, mealy when young, without auricles or oral setae; ligules conspicuous.

Resembles *Indocalamus tessellatus*, but the latter has larger leaves and hairy culm-sheaths which are longer than the internodes.

Native to Japan, where it is widely distributed in the high mountains. It was introduced to Britain in 1889, and flowered sporadically during 1963–1969. It is very hardy, as are most *Sasa* species.

1b. f. nebulosa (*Makino*) *Suzuki*

Sasa paniculata (Schmidt) Makino & Shibata f. *nebulosa* (Makino) Makino & Shibata
S. senanensis (Franch. & Sav.) Rehder f. *nebulosa* (Makino) Rehd.

Distinguished from the typical form by its internodes mottled with purplish brown blotches.

This is the form that is generally grown in gardens all over Europe.

2. Sasa masamuneana (*Makino*) *Chao & Renv.*

Pleioblastus masamuneanus Makino
Arundinaria glabra Nakai
Sasaella glabra (Nakai) Koidz.
S. masamuneana (Makino) Hatsushima & Muroi

Japanese: Kurio-zasa, Genkei-chiku

Culms 1–2 m high, 5–10 mm in diameter, erect; internodes 10–15 cm long, green, glabrous. Culm-sheaths shorter than internodes, glabrous except for the margins, with very small auricles and a few oral setae. Leaf-blades lanceolate or linear-lanceolate, 10–19 cm long, 1.5–2.5 cm wide, glabrous, glaucous beneath; leaf-sheaths purplish on one side, glabrous, with a few short oral setae.

The species is characterized by glabrous culms, culm-sheaths and leaves combined with relatively small leaf-blades.

2a. cv. 'Albostriata'

Sasaella glabra (Nakai) Koidz. f. *albostriata* Muroi

Distinguished from the typical plant by the longitudinal yellowish stripes on its leaf-blades.

Native to Japan. The cultivar 'Albostriata', which is a very beautiful foliage bamboo, has been grown in Britain for several years. It was introduced into the Kew bamboo garden only a couple of years ago and is now growing very well.

3. Sasa borealis (*Hack.*) *Makino & Shibata*

Bambusa borealis Hack.
Arundinaria borealis (Hack.) Makino
Sasamorpha borealis (Hack.) Nakai
S. purpurascens Nakai

Japanese: Suzu-dake

Culms 1–2 m high, 5–10 mm in diameter; internodes relatively short, 8–10 cm long, glabrous, mealy when young. Culm-sheaths longer than internodes, pubescent, without auricles or oral setae; blades lanceolate, erect or spreading. Leaf-blades 2–3 on each branch, 10–20 cm long, 1.5–3 cm wide, glabrous, glaucous beneath; leaf-sheaths glabrous, conspicuously powdery, without auricles or oral setae.

This species is distinguished from other members of the genus by pubescent culm-sheaths longer than the internodes, fairly small leaf-blades, and sheaths without auricles or oral setae.

Native to Japan and Korea. It has recently been introduced into the Kew bamboo garden, although it is grown elsewhere in Britain.

4. Sasa veitchii (*Carr.*) *Rehd.*

Bambusa veitchii Carr.
Arundinaria albo-marginata (Franch. & Sav.) Makino
A. veitchii (Carr.) Brown
Bambusa albo-marginata (Franch. & Sav.) Makino
B. senanensis Franch. & Sav. f. *albo-marginata* Franch. & Sav.
Sasa albo-marginata (Franch. & Sav.) Makino & Shibata

Japanese: Kuma-zasa

Culms 0.5–1.5 m high, 5–7 mm in diameter; internodes short, 5.5–10 cm long, glabrous, green, mealy when young. Culm-sheaths pilose with spreading hairs especially near the base, auricles and oral setae weakly developed or absent; blades small, reflexed. Leaf-blades large, oblong, 10–20 cm long, 4.5–5.5 cm wide, thickly papery, glabrous, with 7–9 pairs of secondary veins and yellowish midrib, usually withered at margins and tip; leaf-sheaths conspicuously glaucous when young, with very short oral setae.

S. *veitchii* is easily distinguished from other species of the genus by its large, oblong leaves with withered margins and tip, and by its pubescent culm-sheaths.

Native to Japan, and commonly cultivated as an ornamental in that country. It was introduced to Britain in 1880, and is now widely distributed in Europe but has not flowered outside Japan.

5. Sasa tsuboiana *Makino*

Japanese: Ibuki-zasa, Tsuboi-zasa.

Culms 1.5–2 m high, 5–10 mm in diameter, branching above; internodes 16–19 cm long, glabrous, with puberulous nodes. Culm-sheaths up to ½ as long as the internodes, glabrous except for the margin and base, the upper sheaths with small auricles and oral setae. Leaf-blades papery, oblong-lanceolate, 15–26 cm long, 3.5–4.5 cm wide, glabrous on both surfaces; leaf-sheaths glabrous with small auricles and short oral setae.

Native to Japan. It was introduced to Britain many years ago, and to the bamboo garden at Kew in 1982 where it grows very well.

6. Sasa takizawana *Makino*

Japanese: Takizawa-zasa

Culms 1–2 m high, 5–7 mm in diameter, erect; internodes 10–14 cm long, glabrous. Culm-sheaths shorter than internodes, pilose with long spreading deciduous hairs, the auricles small with radiate oral setae. Leaf-blades oblong-lanceolate, 16–28 cm long, 3.5–5 cm wide, broadly cuneate at the base, glaucous and puberulous with short soft hairs beneath; auricles conspicuously developed, falcate with long radiate oral setae.

8. Sasa

Native to Japan. It was introduced to the bamboo garden at Kew in 1982 as '*Arundinaria amabilis*', but bears no resemblance to that important medium-sized bamboo from South China. It grows well at Kew.

7. Sasa ramosa (*Makino*) *Makino & Shibata*

Bambusa ramosa Makino
Arundinaria ramosa (Makino) Makino
A. vagans Gamble
Sasaella ramosa (Makino) Makino

Japanese: Azuma-zasa

Culms 1–1.5 m high, 3–8 mm in diameter; internodes 10–15 cm long, light green at first, glabrous, the nodes not prominent. Culm-sheaths shorter than internodes, glabrous, light green at first with purple stripes, without auricles or oral setae. Leaf-blades thin, papery, lanceolate, 10–20 cm long, 2–3 cm wide, sparsely pubescent above, densely white-pubescent beneath, with 5–6 pairs of secondary veins; leaf-sheaths glabrous except for the margins, with a few short deciduous oral setae.

Native to Japan. Introduced to Britain in 1892, and described from a garden plant as *Arundinaria vagans*. Its identity with the native Japanese *S. ramosa* only became apparent when it flowered at Kew in 1981.

This dwarf bamboo is a common garden plant in Britain. It is very hardy and its rhizomes spread with great rapidity. It has been in flower more or less continuously in Britain since 1981.

8. Sasa hayatae *Makino*

Japanese: Miyama-kuma-zasa, Tanzawa-zasa.

Culms 1–1.7 m high, 4–5 mm in diameter, branching above; internodes 11–16 cm long, glabrous. Culm-sheaths conspicuously covered with powder, glabrous, up to ½ the length of the internodes, without auricles or oral setae. Leaf-blades papery, oblong-lanceolate, 15–21 cm long, 2.5–3.5 cm wide, rounded at the base, sparsely pubescent above, densely pubescent beneath; leaf-sheaths without auricles, oral setae sparsely developed.

This species is similar to *Sasa tsuboiana* Makino, but is distinguished by its small pubescent leaves.

Native to Japan. It was introduced to Kew Gardens, originally under the mistaken name of '*Sasa matsudae*', and flowered in 1965.

9. Indocalamus *Nakai*

Shrubs; rhizomes monopodial or with some sympodial branching interspersed among the runners; internodes terete without a groove, the nodes usually not prominent; branch-complement usually 1, subequal to culm in diameter, exceptionally 3 at upper nodes of culm. Culm-sheaths persistent or tardily deciduous. Leaf-blades very large with many secondary veins, tessellate. Inflorescence paniculate. Spikelets several-flowered with pedicels; glumes 2; stamens 3; stigmas 2.

Outwardly similar to *Sasa*, differing mainly in the number of stamens and stigmas.

About 15 species. China.

Indocalamus tessellatus *(Munro) Keng f.*

Bambusa tessellata Munro
Sasa tessellata (Munro) Makino & Shibata
Sasamorpha tessellata (Munro) Nakai

Chinese: Ruo zhu

Culms 2–2.5 m high, 1–1.5 cm in diameter, thick-walled; internodes short, about 10–15 cm long, light green with heavy bloom. Culm-sheaths longer than internodes, light green at first, fading to brownish yellow, sparsely hirsute, without auricles or oral setae; blades lanceolate. Leaf-blades large, 25–50 cm long, 4–8 cm wide or more, glaucous and sometimes with a line of hairs down one side of the lower midrib beneath, with conspicuous yellow midrib and 13–18 pairs of secondary veins.

Characterized by its short internodes, long culm-sheaths and very large leaf-blades with numerous secondary veins.

Native to China. Introduced to Britain in 1845. One culm flowered briefly at Kew Gardens in 1984. More extensive flowering was observed in southern France in 1988.

In China the leaves are used as a wrapping for pyramid-shaped dumplings of glutinous rice.

10. Chusquea *Kunth*

Trees, shrubs or climbers; rhizomes sympodial, rarely monopodial. Culms solid, rarely with an irregular lumen, terete or cylindrical; branch-complement many, slender, developed from separate primary buds, sometimes the central bud producing a strongly dominant branch. Culm-sheaths persistent. Inflorescence usually an open panicle, rarely racemose or capitate. Spikelet typically 1-flowered without rhachilla-extension, rarely 2-flowered; glumes 4; lodicules 3; stamens 3; stigmas 2.

The genus is characterized by its solid culms, persistent culm-sheaths, and many slender branches developed from separate buds.

About 100 species. Central and South America. It is a large group whose species are difficult to distinguish. Two species have been generally introduced to Britain.

Key to species of *Chusquea*

Leaf-blades 4–7 mm wide, usually with 2 secondary veins on each side and very fine tessellation **1. C. culeou**
Leaf-blades 1–1.5 cm wide usually with 3–4 secondary veins on each side and inconspicuous tessellation **2. C. quila**

1. Chusquea culeou *Desv.*

Chusquea andina Philippi
C. breviglumis Philippi

Culms 4–6 m high, 2–4 cm in diameter, erect, tapering; internodes 10–15 cm long, olive-green at first, yellowish or yellow-brown at maturity; branch-complement many, very slender. Culm-sheaths longer than internodes, purplish at first, quickly fading to white, densely puberulous above, without auricles or oral setae; blades very narrow. Leaf-blades small, 4–10 cm long, 4–7 mm wide, midrib prominent, secondary veins usually 2 on each side, very finely tessellate, puberulous beneath; leaf-sheaths with long conspicuous ligules, without auricles or oral setae.

Native to Chile. Introduced to Britain in 1890, and now grown in many parts of Europe.

This species is very hardy, and able to withstand considerable periods of low temperature.

2. Chusquea quila *Kunth*

Vernacular: Quila, Kili, Keelee

Culms often climbing up to 10–15 m high; internodes 15–22 cm long, glabrous, smooth; branch-complement many, usually with one dominant branch. Leaf-blades lanceolate or linear-lanceolate, 7–15 cm long, (0.7–)1–1.5 cm wide, pubescent beneath, midrib prominent, secondary veins usually 3 or 4 on each side, inconspicuously tessellate; leaf-sheaths pubescent at apex.

Native to Chile. Introduced to Kew Gardens in 1983.

11. Sinarundinaria *Nakai*

Burmacalamus Keng f.
Butania Keng f.
Chimonocalamus Hsueh & Yi
Drepanostachyum Keng f.
Otatea (McClure & Smith) Cald. & Soderstrom
Yushania Keng f.

Small or medium bamboos; rhizomes sympodial with short or long neck, forming dense or spaced clumps. Culms erect, terete, sometimes with thorns at nodes; branch-complement 3–many, arising from one bud. Culm-sheaths tardily deciduous. Inflorescence paniculate or racemose, exserted. Spikelets 1–several-flowered with pedicels; glumes 2; stamens 3; stigmas 3 or 2.

The inflorescence and spikelets are similar to *Arundinaria*, but it is easily distinguished by its sympodial rhizomes and different habit. It is also similar to *Thamnocalamus* in the vegetative state, but the inflorescence types of these two genera are quite different.

About 50 species, Asia, South America, Africa and Madagascar. 3 species have been introduced to Britain.

Key to species of *Sinarundinaria*

1 Leaf-blades without transverse veinlets; ligule of culm-sheaths 5–12 mm long; culm-sheaths sparsely setose, attenuate towards the apex **1. S. falcata**
1 Leaf-blades tessellate with conspicuous transverse veinlets; ligule of culm-sheaths only 1–2 mm long: 2
  2 Culm-sheaths densely setose, without auricles or oral setae; leaf-blades 3.5–7 cm long; leaf-sheaths with straight oral setae **2. S. nitida**
  2 Culm-sheaths glabrous, with auricles and oral setae; leaf-blades 7–12 cm long; leaf-sheaths with radiate oral setae **3. S. anceps**

1. Sinarundinaria falcata (*Nees*) *Chao & Renv.*

Arundinaria falcata Nees
Chimonobambusa falcata (Nees) Nakai
Drepanostachyum falcatum (Nees) Keng f.
Thamnocalamus falcatus (Nees) E.G. Camus

Culms 4–6 m high, 2–2.5 cm in diameter; internodes 20–25 cm long, glabrous, greyish green, densely mealy when young, the sheath-scars thick and much swollen; leafy branchlets very long, wiry. Culm-sheaths green, as long as or longer than internodes, sparsely setose, finely ciliate on the margins, tapering to a narrow point, auricles and oral setae lacking; ligule conspicuous, 5–12 mm long; blade narrow, 2.5–4 cm long, reflexed. Leaf-blades linear-lanceolate, 11–21 cm long, 1–2.4 cm wide, glabrous, not visibly tessellate; leaf-sheaths without auricles or oral setae; ligules markedly elongated, acute at apex.

This species is characterized by long tapering culm-sheaths with very long ligules and by leaf-blades without visible tessellation. It is believed that most British specimens are *S. falcata*. However, *S. hookeriana* (Munro) Chao & Renv., another Himalayan bamboo with 1-flowered (instead of 2-flowered) spikelets and completely glabrous culm-sheaths, has also been introduced, and it is extremely difficult to distinguish between the two species in the vegetative state.

Native to Himalaya. This is the well-known low level species of NW. Himalaya, always found in the undergrowth of forests in that area. It was introduced to Britain by Nees about 1870, and has flowered several times.

It is an attractive, but not very hardy, species.

2. Sinarundinaria nitida (*Stapf*) *Nakai*

Arundinaria nitida Stapf

English: Fountain bamboo
Chinese: Jian zhu (Arrow bamboo)

Culms 4–5 m high, 1–1.5 cm in diameter, arching above; internodes 20 cm long, green and densely mealy when young, quickly turning to purple, glabrous, nodes not prominent, sheath-scars more or less swollen, unbranched till second year; leafy branchlets numerous, usually purple. Culm-sheaths tardily deciduous, shorter than internodes, light green with purplish tint when fresh, densely purple-pubescent, tapered towards the apex, without auricles or oral setae; blades linear. Leaf-blades 4.5–7 cm long, 5–10 mm wide, glabrous; leaf-sheaths with very short oral setae.

It may be confused with *Thamnocalamus spathaceus*, but is distinguished by its purple culms, purple-pubescent culm-sheaths with tapered apex and smaller leaf-blades.

Native to W. China. Seeds collected by M. Berezovski in S. Kansu province were sent to the Imperial Botanic Gardens at St. Petersburg, now Leningrad, in 1886, and a portion of those seeds was given to Kew Gardens in 1889. The species has never been recorded as having flowered in Europe.

This bamboo is very hardy and rather graceful as a decorative plant in gardens. It is one of the bamboos which pandas very much like to eat.

3. Sinarundinaria anceps (*Mitf.*) *Chao & Renv.*

Arundinaria anceps Mitf.
A. jaunsarensis Gamble

11. Sinarundinaria

Yushania anceps (Mitf.) Lin
Y. jaunsarensis (Gamble) Yi

Culms scattered (rhizome sympodial but with a neck up to 1 m long), 3–6 m high, 1–2 cm in diameter, erect in habit at first, arching when mature; internodes greenish brown when mature, glabrous, mealy below the nodes; leafy branchlets numerous, purplish. Culm-sheaths shorter than internodes, glabrous, ciliate on the margins, mealy below the nodes, with falcate auricles and stiff oral setae; blades small, subulate; ligules short. Leaf-blades thin, linear-lanceolate, 6–12 cm long, 1 cm wide, sparsely pilose when young, with 4–5 secondary veins on each side, tessellate; leaf-sheaths with conspicuous auricles and radiate oral setae.

The species is characterized by its well-spaced culms, glabrous auriculate culm-sheaths, and leaf-sheaths with conspicuous auricles and radiate oral setae.

Native to India. Introduced to Britain in 1865. In the spring of 1896 plants cultivated in Britain were given the name *A. anceps*. In the same year native Indian plants were described as *A. jaunsarensis*; the precise date of publication is uncertain, but it appears to have been in late summer. The earlier name is therefore *A. anceps*.

This bamboo has flowered several times in Britain, firstly in 1910–1911 and latterly in 1980–1981.

12. **Thamnocalamus** *Munro*

Fargesia Franch.
Himalayacalamus Keng f.

Small trees; rhizomes sympodial with very short neck. Culms rather close together, the internodes terete without a groove; branch-complement 3–many, arising from one bud. Culm-sheaths deciduous. Inflorescence comprising several racemes, these partly enveloped by large spathaceous bracts. Spikelets 1–several-flowered; glumes 2; stamens 3; stigmas 3.

The genus is distinguished by its spathaceous inflorescence, and is most difficult to separate from *Sinarundinaria* in the vegetative state.

Key to species of *Thamnocalamus*

1 Leaf-blades without visible transverse veinlets   **1. T. falconeri**
1 Leaf-blades with conspicuous transverse veinlets: 2
  2 Culm-sheaths glabrous except for the margins: 3
    3 Culm-sheaths with long ciliate ligules; leaf-sheaths with dark coloured auricles and several long oral setae, the ligule long and dark coloured
         2. T. spathiflorus
    3 Culm-sheaths with very short glabrous ligules; leaf-sheaths without auricles but with 1–3 light coloured oral setae, the ligule short and light coloured
         3. T. spathaceus
  2 Culm-sheaths more or less pubescent: 4
    4 Culms more or less zigzag, yellowish when mature; culm-sheaths deciduous, with long bulbous-based bristles   **4. T. aristatus**
    4 Culms straight, purplish when mature; culm-sheaths persistent, with short soft hairs   **5. T. tessellatus**

1. Thamnocalamus falconeri *Munro*

Arundinaria falconeri (Munro) Benth. & Hook.f.
A. nobilis Mitf.
Himalayacalamus falconeri (Munro) Keng f.

Indian: Pummon, Pao mung

Culms 7–10(–20) m high, 3–8 cm in diameter, the outer usually spreading; internodes 20–40 cm long, olive-green at first, yellowish when mature, purplish on the slightly raised nodes; branch-complement numerous, slender. Culm-sheaths deciduous, light crimson or purplish, glabrous except for the ciliate margins, without auricles or oral setae; ligules short, 1–2 mm long. Leaf-blades small and thin, 6–12 cm long, 6–12 mm wide, light green, not or at most obscurely tessellate; leaf-sheaths without auricles or oral setae.

The species is easily distinguished from other members of the genus by the absence of regular transverse veinlets. The leaf-blades resemble those of *Sinarundinaria falcata* but the latter has gradually and concavely attenuate culm-sheaths and very long ligules.

Native to NE. Himalaya, Nepal, Bhutan and India. It was introduced to Britain in 1847 by Mr Madden, who sent large quantities of seed to Kew, whence they were distributed through Europe. It flowered for the first time in Britain in 1875–1877. All the parent plants subsequently died, but a new generation was raised from seed. Further flowering has occured in 1903–1908, 1929–1932 and 1964–1969.

This bamboo is not fully hardy, but can be grown well in warmer and more sheltered areas. The culms are very pliable and are used extensively for fishing rods and woven baskets.

2. Thamnocalamus spathiflorus *(Trin.) Munro*

Arundinaria spathiflora Trin.

Culms 4–7 m high, 2–2.5 cm in diameter, erect with a tendency to zigzag; internodes glaucous green with white bloom at first, later pinkish purple on the exposed side, the nodes with prominent sheath-scars. Culm-sheaths deciduous, often asymmetric, coriaceous, glabrous except for the upper margin, without auricles but with a few short oral setae; ligules long, ciliate at apex. Leaf-blades thin, 9–13 cm long, 1–1.3 cm wide, regularly tessellate beneath; leaf-sheaths with dark-coloured auricles and oral setae; ligules long and dark-coloured.

The species is characterized by its glabrous culm-sheaths, and by its leaf-sheaths with dark-coloured auricles, oral setae, and ligules.

Native to NW. Himalaya, Nepal and India. It was introduced to Britain in 1886 and flowered in a garden at Chobham, Surrey in 1970.

It is an extremely handsome and elegant bamboo, but is not very hardy, and should only be grown in more sheltered areas away from cold winds.

3. Thamnocalamus spathaceus *(Franch.) Soderstrom*

Fargesia spathacea Franch.
Arundinaria murielae Gamble
A. spathacea (Franch.) D. McClint.
Sinarundinaria murielae (Gamble) Nakai

English: Umbrella bamboo

Chinese: Xian zhu (Thin bamboo)

Culms 3–4 m high, 1–1.5 cm in diameter; internodes about 20 cm long, light green with white bloom at first, yellow when mature. Culm-sheaths deciduous, light green at first but quickly turning to straw-coloured, glabrous except for the ciliate margins, rounded at apex, without auricles or oral setae; ligules very short; blades reflexed. Leaf-blades small, 6–10 cm long, 8–13 mm wide, glaucous beneath, finely tessellate; leaf-sheaths with 1–3 white oral setae.

This species is similar to *T. spathiflorus*, but is distinguished by its few pale oral setae, short pallid ligules, and quite different spathaceous bracts. It is much confused with *Sinarundinaria nitida*, the main difference being that *S. nitida* has purple culms and purple-pubescent culm-sheaths.

Native to W. China. First taken from Hubei province of China to the Arnold Arboretum, U.S.A., by Wilson in 1907, and thence to Kew in 1913. It is now grown in many parts of Europe, and flowered in 1978–1980.

T. spathaceus is a favourite food of pandas, whose native distribution coincides with that of the bamboo. It is very hardy and grows well in Europe.

4. Thamnocalamus aristatus (*Gamble*) E.G. Camus

Arundinaria aristata Gamble

Vernacular: Bhebkam (Bhutia), Babain (Lepcha)

Culms 5–8 m high, 2–3(–6) cm in diameter, sometimes more or less zigzag; internodes 15–25 cm long, glaucous green at first, ageing to brownish green, yellowish when mature, with reddish branches. Culm-sheaths deciduous, sparsely hairy, with inconspicuous auricles; blades small, subulate. Leaf-blades 8–11 cm long, about 1 cm wide, glaucous beneath; leaf-sheaths with a few light-coloured oral setae.

The species is similar to *T. spathiflorus* in many respects, but the latter has glabrous culm-sheaths, leaf-sheaths with dark-coloured ligules and oral setae, and quite narrow spathaceous bracts.

Native to NE. Himalaya, Sikkim and Bhutan. It was introduced to Britain at the end of last century, and is not uncommon in Europe. It flowered at Kew in 1950.

5. Thamnocalamus tessellatus (*Nees*) Soderstrom & Ellis

Nastus tessellatus Nees
Arundinaria tessellata (Nees) Munro

Afrikaans: Bergbamboes

Culms 4 m high (6–7 m in native habitat), 1 cm in diameter, thin-walled, erect. Internodes 20–23 cm long, light green at first, purplish when mature; branch-complement numerous, of short purplish branches. Culm-sheaths persistent, longer than internodes, whitish at first, with soft short hairs, the auricles and oral setae developed. Leaf-blades 5–10(–14) cm long, 0.6–1.1(–1.5) cm wide, conspicuously tessellate; leaf-sheaths with auricles and oral setae.

The species is characterized by its persistent pubescent culm-sheaths, visibly tessellate leaf-blades and very short branches.

Native to South Africa. It has been introduced to Britain, but is uncommon.

13. Dendrocalamus *Nees*

Neosinocalamus Keng f.
Sinocalamus McClure

Trees; rhizomes sympodial, forming clumps. Culm-internodes terete without a groove; branch-complement many, 1 dominant, none forming thorns. Culm-sheaths thick and stiff, usually without auricles or almost so, rarely auricles conspicuous; blades often much narrower than apex of sheaths. Leaf-blades usually very large. Inflorescence comprising a globose mass of many spikelets condensed around the nodes. Spikelets 1–6-flowered, the rhachilla abbreviated and obscure; glumes 2–3; lemmas increasing in length upwards; lodicules usually absent, rarely 2–3; stamens 6, free or united into a flimsy tube; stigma usually 1 on an elongated style.

According to the traditional system of bamboo classification, *Dendrocalamus* and *Bambusa* belong to different subtribes, but in fact they are very similar to each other in both morphology and anatomy. They are difficult to distinguish without flowers, but *Dendrocalamus* usually has larger leaf-blades and culm-sheaths without auricles.

About 35 species. China to South Asia. Only one species is grown in Britain.

Dendrocalamus latiflorus *Munro*

Sinocalamus latiflorus (Munro) McClure

Chinese: Tian zhu (Sweet bamboo), Ma zhu
Japanese: Ma-chiku

Culms up to 20 m high, 10–20 cm in diameter, thick-walled; internodes up to 50–70 cm long, the nodes rather prominent. Culm-sheaths coriaceous, hard and brittle, yellowish green at first, light brown when dry, densely covered with dull brown short appressed hairs, rounded at apex, auricles not developed; blades small, reflexed. Leaf-blades 20–40 cm long, 3–8 cm wide, puberulous beneath; leaf-sheaths without oral setae.

Native to S. China. This bamboo is widely planted in Taiwan and S. China for its excellent edible shoots. They are large, relatively solid, and unusually free from any unpleasant taste when raw, making them an important commercial crop.

It is not hardy, but its dimensions make if an impressive feature of the largest greenhouses in some British botanical gardens.

14. Bambusa *Schreb.*

Leleba Nakai

Trees or shrubs, rarely scramblers; rhizomes sympodial, forming clumps; internodes terete without a groove; branch-complement many, originating from one bud, sometimes with branch-thorns at the node. Culm-sheaths thick and stiff, usually with auricles; blades broad, triangular. Leaf-blades often small or medium-sized. Inflorescence comprising a tuft of 1–many spikelets condensed about a node. Spikelets sessile, supported by small bracts, 2–many-flowered, the rhachilla distinct; glumes 1–3; lemmas subequal; lodicules usually 3; stamens 6, free; stigmas usually 3, on a short or elongated style.

14. Bambusa

Bambusa is the largest genus of bamboos, comprising about 120 species. Tropical Asia and America. Two species have been introduced to Britain.

Key to species of *Bambusa*

1 Culm-sheaths densely brown-hirsute, the auricles well developed; culms 5–10 cm in diameter **1. B. vulgaris**
1 Culm-sheaths glabrous, without auricles; culms less than 3 cm in diameter: 2
  2 Culms hollow, 3–5 m high; leaf-blades 9–16 cm × 7–15 mm: 3
    3 Culms uniformly green: 4
      4 Leaf-blades uniformly green **2a. B. glaucescens** var. **glaucescens**
      4 Leaf-blades green with white stripes **2b. B. glaucescens** cv. Silverstripe
    3 Culms yellow with green stripes **2c. B. glaucescens** cv. Alphonse-Karr
  2 Culms solid, less than 2 m high; leaf-blades 4–7 cm × 5–7 mm, markedly distichous **2d. B. glaucescens** var. **riviereorum**

1. Bambusa vulgaris *Wendl.*

Leleba vulgaris (Wendl.) Nakai

English: Common bamboo
Chinese: Longtou zhu (Dragon's head bamboo), Taishan zhu
Japanese: Daisan-chiku

Culms 5–7(–25) m high, 5–10 cm in diameter, with an open-clump habit; internodes green at first, yellow later. Culm-sheaths densely hirsute with short appressed stiff brown hairs, most of which gradually fall off, the auricles conspicuous, falcate, fringed by stiff oral setae; blades, on the lower sheaths especially, much narrower than the apex of the sheath. Leaf-blades 15–25 cm long, 1–3 cm wide, densely and minutely pubescent beneath.

This species is characterized by its open-clump habit, culm-sheaths covered in dense brown hairs and symmetric sheath auricles.

Its original habitat is not definitely known, but is generally believed to have been India. It is one of the most widely grown of all bamboos in tropical and subtropical regions of both hemispheres, providing a convenient source of village timber for general constructional work.

It is not hardy in Britain, but is occasionally grown as a large hothouse plant.

2. Bambusa glaucescens (*Willd.*) *Munro*

Ludolfia glaucescens Willd.
Bambusa multiplex (Lour.) Raeuschel
B. nana Roxb.
Leleba multiplex (Lour.) Nakai

English: Hedge bamboo, Oriental Hedge bamboo
Chinese: Xiao-shun zhu (Filial obedience bamboo)
Japanese: Horai-chiku

2a. var. glaucescens

Culms 3–5 m high, 2–3 cm in diameter, thin-walled, arched; internodes green or yellowish green, pubescent at first, soon glabrous. Culm-sheaths soon glabrous, stiff, without auricles or oral setae; blades large, triangular, decurrent;

ligules very short, 1–1.5 mm long. Leaf-blades 4–8 per branch, usually 9–16 cm long, 7–15 mm wide, minutely pubescent beneath; leaf-sheaths with auricles and oral setae.

This species is easily distinguished from other members of the genus by its smaller size, culm-sheaths without auricles and large sheath-blades.

Native to S. China, but long cultivated in the East Indies, Malay Archipelago and Japan. It was introduced to Europe in the last century, but is rarely grown.

B. glaucescens is the commonest species of clump-type bamboo cultivated in China, Japan and southern California. It is the hardiest member of the genus, and is reported to have withstood temperatures as low as −9°C.

Hedge bamboo is one of the most variable of all species in cultivation. There are eight distinctive cultivars or varieties, three of which have been grown in Britain.

2b. cv. 'Silverstripe'

Bambusa multiplex f. *variegata* Nakai
B. multiplex f. *vittate-argentea* Nakai
B. nana var. *variegata* E.G. Camus
B. vittate-argentea E.G. Camus in synon.

Differs from the typical plant in having some or most of its leaf-blades striped with white; also the culm internodes usually bear one or two threadlike whitish stripes. It is sometimes mistakenly labelled '*B. argenteo-striata*' in gardens.

2c. cv. 'Alphonse Karr'

Bambusa alphonse-karrii Satow
B. multiplex f. *alphonse-karrii* (Satow) Nakai
B. nana var. *alphonse-karrii* (Satow) H. de Lehaie
Leleba multiplex f. *alphonse-karrii* (Satow) Nakai
Bambusa multiplex cv. Alphonse Karr

English: Alphonse Karr bamboo, Golden-striped hedge bamboo
Chinese: Hua Xiau-shun zhu (Coloured filial obedience bamboo)
Japanese: Ho-o-chiku

Much like the typical plant in stature and foliage characters, but the culms and branches are golden yellow with longitudinal green stripes of different widths.

This is a very beautiful and decorative bamboo. It is rare in Britain, though adapted to locations where winter temperatures do not often fall below −8°C.

2d. var. **riviereorum** (*M. Maire*) *Chia & Fung*

B. multiplex var. *riviereorum* M. Maire

English: Chinese goddess bamboo, Riviere hedge bamboo
Chinese: Guan-yun zhu (Goddess bamboo)

Culms up to 2 m high, solid. Leaf-blades small, fern-like, 4–7 cm long, 5–7 mm wide, markedly distichous.

Sometimes mistaken for cv. 'Fernleaf'; the latter has similar leaves but taller hollow culms.

In China this beautiful bamboo is often used for miniature landscapes in tubs. It is rarely planted in Britain, though it tolerates −8°C.

INDEX OF VERNACULAR NAMES

Index

INDEX OF BOTANICAL NAMES

Printed in Great Britain by
Whitstable Litho Printers Ltd., Whitstable, Kent